AF830029

World Soils Book Series

Series Editor

Alfred E. Hartemink
Department of Soil Science, FD Hole Soils Laboratory
University of Wisconsin–Madison
Madison, WI
USA

The World Soils Book Series publishes peer-reviewed books on the soils of a particular country. They include sections on soil research history, climate, geology, geomorphology, major soil types, soil maps, soil properties, soil classification, soil fertility, land use and vegetation, soil management, soils and humans, soils and industry, future soil issues. The books summarize what is known about the soils in a particular country in a concise and highly reader-friendly way. The series contains both single and multi-authored books as well as edited volumes. There is additional scope for regional studies within the series, particularly when covering large land masses (for example, The Soils of Texas, The Soils of California), however, these will be assessed on an individual basis.

More information about this series at http://www.springer.com/series/8915

Hamdi A. Zurqani
Editor

The Soils of Libya

Springer

Editor
Hamdi A. Zurqani
Department of Soil and Water Sciences
University of Tripoli
Tripoli, Libya

ISSN 2211-1255 ISSN 2211-1263 (electronic)
World Soils Book Series
ISBN 978-3-030-66367-4 ISBN 978-3-030-66368-1 (eBook)
https://doi.org/10.1007/978-3-030-66368-1

This Springer imprint is published by the registered company Springer Nature Switzerland AG
The registered company address is: Gewerbestrasse 11, 6330 Cham, Switzerland

Preface

Soil is the source of life, and there can be no life without it. Each soil has had its own history and its present condition is due to the influences of many different factors (e.g. climate, biota, parent material, and relief or topography). The total land area in Libya is about 180 million hectares, and its population is about 6 million, and it is located in the arid and semi-arid regions of the world. Libya's climate is dominantly influenced by both the Mediterranean Sea in the northern regions and the desert in the southern regions. Thus, most of the soils are undeveloped or partially developed. This book helps the reader to understand the soil pedodiversity in Libya. The book's chapters provide adequate information about Libyan soils. Chapter 1 gives an introduction to Libyan soils and discusses soil-related challenges and potential solutions. Chapter 2 covers the history of soil mapping and research in Libya. Chapter 3 discusses the soil formation factors and their impact on the soil formation processes prevailing in different regions across the country. Chapter 4 discussed the most common soil classification systems used in the country and reviewed the main properties of the Libyan soils. Chapter 5 illustrates the constraints and limiting factors that negatively affect agricultural activities across the country. Land cover/land use and vegetation distribution in Libya are described in Chap. 6. Chapter 7 presents the status of soil biology, the corresponding related research activities, and the other biological properties of the Libyan soils. The final chapter (Chap. 8) discusses land degradation and desertification in Libya, emphasizing the main causes, impacts of the phenomena, and efforts to combat it. This book has a valuable contribution to many distinguished scientists from various universities and institutions in and outside of Libya.

Tripoli, Libya Hamdi A. Zurqani

Acknowledgments

First of all, I would like to acknowledge Dr. A. E. Hartemink, Editor of the Springer's World Soils Book Series, Dr. Robert Doe, and all the staff of Springer that kindly assisted us in our multiple queries during the elaboration of this book. Special thanks are owed to Prof. Khaled R. Ben Mohamed, for his guidance and suggestions which helped bring this book into existence. I would also like to thank all the authors for their dedication, effort, and enthusiasm. Finally, I wish to express my appreciation to all soil scientists who have labored to bring an understanding of the soil resources in Libya and their pivotal role in agricultural productivity.

February 2021

Editor
Dr. Hamdi A. Zurqani
Department of Soil and Water Sciences
Faculty of Agriculture, University of Tripoli
Tripoli, Libya

The Soils of Libya

Highlights

- The first book with a comprehensive characterization of the Libyan soils published in English
- Provides an overview of the challenges that the soils of Libya face in relation to agriculture and the environment
- Written by distinguished Libyan soil experts
- A valuable resource for anyone interested in soils and agriculture in Libya or any other country with similar climatic conditions.

About This Book

This is the first comprehensive book on the Libyan soils published in English. The book contains eight chapters that address the history of soil mapping, education, soil geographical distribution, classification, properties, functions, agricultural use, and environmental aspects including land degradation and desertification in Libya. This comprehensive and richly illustrated book offers a valuable resource for anyone interested in soils and agriculture in Libya or any other country with similar climatic conditions.

Contents

Editor and Contributors

About the Editor

Dr. Hamdi A. Zurqani is one of the faculty members at the Department of Soil and Water Sciences, Faculty of Agriculture, University of Tripoli, Tripoli, Libya. Dr. Zurqani is a recognized expert as a result of his internationally acclaimed work in the areas of environmental information science, remote sensing, land evaluation, sustainability, pedology, and soil science education. He has conducted research across the world, including the United States of America and Africa. Dr. Zurqani is a distinguished soil scientist with a wide range of scientific and working experiences in Libya and abroad. He received his M.Sc. (2010) from the University of Tripoli, Tripoli, Libya and Ph.D. (2019) from Clemson University, Clemson, SC, USA. His major research and teaching activities at the University of Tripoli have focused mainly on soil genesis and classification and the environmental information sciences (Remote Sensing and Geographic Information System). He has published broadly in many journals (e.g., Nature "Scientific Reports", Geoderma; International Journal of Applied Earth Observation and Geoinformation; Journal of Geological Sciences; Land; Frontiers in Environmental Science; Communications in Soil Science and Plant Analysis; and others). Dr. Zurqani is a member of the Editorial Board for Remote Sensing (MDPI) Journal, counseling outcome and research evaluation. He also was appointed to serve as a Guest Editor for the Special Issue "Applications of Remote Sensing in Earth Observation and Geo-Information Science". In addition, Dr. Zurqani conducted peer-review for many journals including Journal of Environmental Informatics, Applied Sciences, SN Applied Sciences, Remote Sensing, Heliyon, Geosciences, Land, Water, Agronomy, Agriculture, Sustainability, Arid Land Research and Management, International Journal of Environmental Research and Public Health, Natural Hazards, and Conference of the Arabian Journal of Geosciences. He is also one of the authors of the lab manual entitled, "GIS Exercises for Natural Resource Management". Dr. Zurqani has been the recipient of numerous awards and honors: Recipient of Douglas R. Phillips Award for Graduate Students, Department of Forestry and Environmental Conservation, Clemson University,

April 12, 2019; the First Place Best Judged Poster (CAFLS) at the GRADS 2019: Clemson Student Research Forum on April 4, 2019; the Second Place Poster at the 11th Clemson Biological Sciences Annual Student Symposium, April 6, 2019; the Second Place Best Judged Poster at the Clemson Student Research Forum on April 4, 2018; and the Third Place Poster at the 9th Clemson Biological Sciences Annual Student Symposium, February 25, 2017. Dr. Zurqani conducts cutting-edge research in the field of environmental information science, remote sensing, land use management/planning, change detection of landscape degradation, and geographic information system (GIS) models. He has focused on his research efforts on the development of new technologies in the field of environmental information sciences, geo-intelligence (advanced geo-information science and earth observation, machine and deep learning, and big data analytics), remote sensing, land evaluation, pedology, land use management/ planning, monitoring and evaluating sustainable land management, change detection of landscape degradation, and geographic information system models.

Contributors

Azalarib S. Ali Faculty of Agriculture, Department of Soil and Water Sciences, University of Tripoli, Tripoli, Libya

Asma Yousef Alnajjar Faculty of Agriculture, Department of Soil and Water Sciences, University of Tripoli, Tripoli, Libya

Khaled R. Ben Mahmoud Faculty of Agriculture, Department of Soil and Water Sciences, University of Tripoli, Tripoli, Libya

Merfat Taher Ben Mahmoud Faculty of Agriculture, Department of Soil and Water Sciences, University of Tripoli, Tripoli, Libya

Mukhtar M. Elaalem Faculty of Agriculture, Department of Soil and Water Sciences, University of Tripoli, Tripoli, Libya

Azzeddin R. Elhawej Faculty of Agriculture, Department of Soil and Water Sciences, University of Tripoli, Tripoli, Libya

Eman Ali Ferjani Faculty of Agriculture, Department of Soil and Water Sciences, University of Tripoli, Tripoli, Libya

Basher A. Nwer Faculty of Agriculture, Department of Soil and Water Sciences, University of Tripoli, Tripoli, Libya

Hamdi A. Zurqani Faculty of Agriculture, Department of Soil and Water Sciences, University of Tripoli, Tripoli, Libya

1 Introduction

Hamdi A. Zurqani

Abstract

Soil is an essential and nonrenewable natural resource from which most of the global food is produced. Each soil has had its own history and its present conditions due to the influences of many different factors. This chapter presents an introduction about the soils of Libya and demonstrates the challenges that the country is currently facing as a result of soil erosions, salinization, pollution, and climate change while proposing the potential solutions to these challenges to improve food security in the country. Libya covers an area of about 180 million hectares and located in the arid and semiarid regions of the world. Thus, most of the soils are undeveloped or partially developed. Dry climatic conditions in Libya resulted in high accumulation of calcium carbonate in soils, with the presence of gypsum in some areas as well as salt accumulation which leads to naturally saline soils. Most agricultural activities are mainly located on a narrow strip along the Mediterranean coast that receives the highest rainfall in the country (up to 600 mm annually), and in low mountains and scattered oases in the desert.

Keywords

Soils of Libya • Arid and semiarid soils • Soil information • Soil erosion • Salinization • Land degradation

H. A. Zurqani (✉)
Department of Soil and Water Sciences, Faculty of Agriculture, University of Tripoli, Tripoli, Libya
e-mail: h.zurqani@uot.edu.ly

H. A. Zurqani (ed.), *The Soils of Libya*, World Soils Book Series,
https://doi.org/10.1007/978-3-030-66368-1_1

1.1 Introduction

The State of Libya is a country in the Maghreb region of North Africa, lies along the southern coast of the Mediterranean, approximately between latitude 18° and 33° north and 9° and 25° east. It is the fourth largest country in Africa with an area of approximately 1,759,540 km^2 and a population of about 6 million residents (Zurqani et al. 2019). Libya is bounded by the Mediterranean Sea to the north, Egypt to the east, Sudan to the south-east, Chad and Niger to the south, Algeria to the west, and Tunisia to the north-west (Fig. 1.1).

Soils in the literal sense of the word are a finite resource in Libya where most of the country lies in "**The Sahara Desert**", which is estimated at about 95% of the total area. Most of the country lies within the arid and semiarid regions except for a small area in Al Jabal Al Akhdar (English: The Green Mountain) which is ranged from semi-arid to sub-humid (Fig. 1.1). The soils of Libya and their characteristics are largely influenced to the great extent by factors of soil formation and local conditions in which these soils were formed. Generally, aridity is the main influence on such soils, where they are frequently undeveloped or partially developed. Thus, Libyan soils are slightly or moderately weathered and may include one diagnostic horizon or a few diagnostic horizons that are characteristic of the arid and semiarid regions (Zurqani et al. 2018). Only about 2% of the country is arable land, with the rest comprising rocky outcrops and loose surface materials. Land degradation and desertification are the major soil threats facing the agricultural development of the region. The majority of Libyan soils are sandy in texture and tend to have low water-holding capacities and high infiltration rates and are prone to loss of water through rapid drainage below the crop roots with a concomitant loss of water-soluble plant nutrients from the soil (Ben Mahmoud 1995). In terms of the soil's spatial distribution, the diversity of the Libyan soils is low. Rich alluvial soils are found in the coastal deltas and valleys of

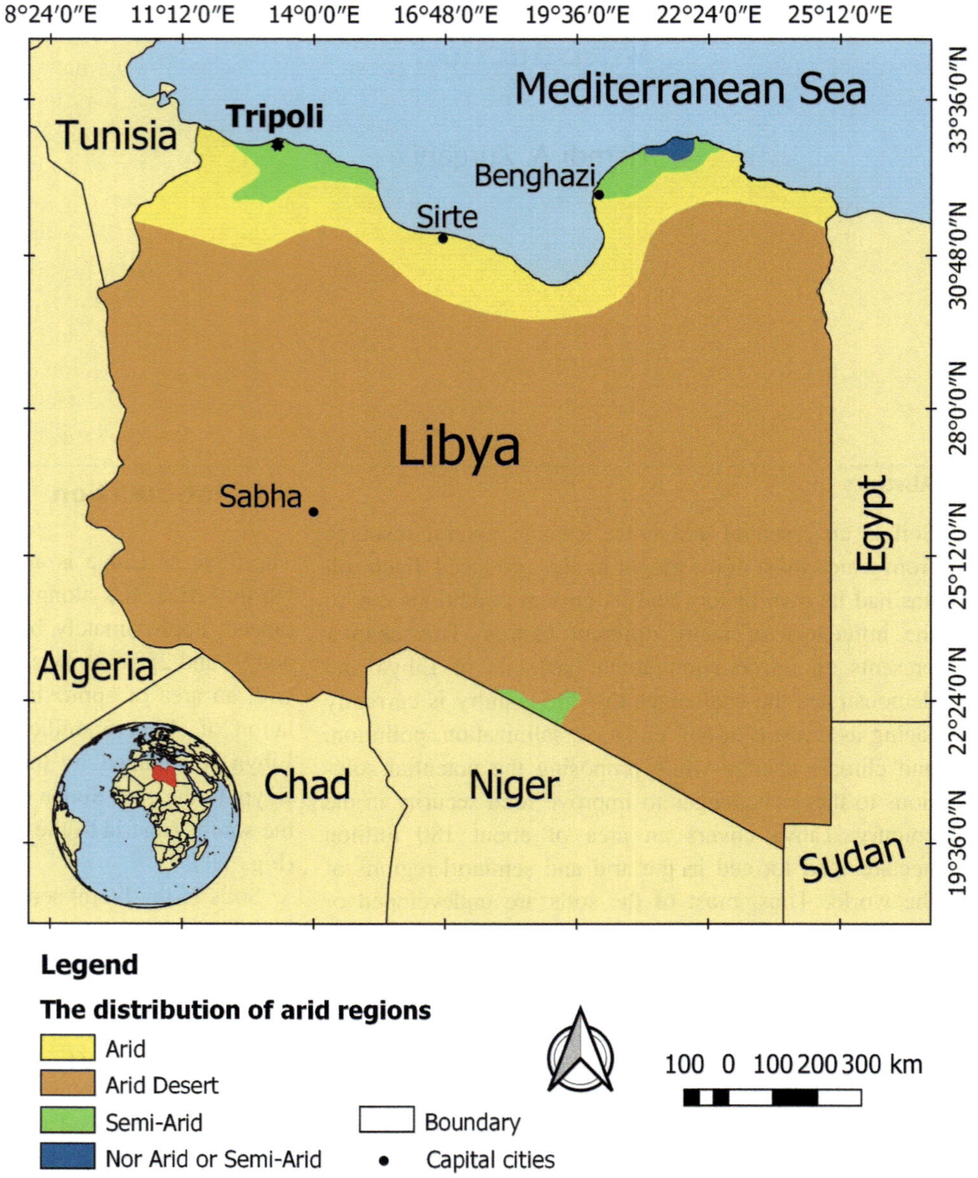

Fig. 1.1 Map of Libya with capital cities and the distribution of arid regions adapted from (Ben Mahmoud 1995)

large wadis. The most fertile soils in Libya occur in the Benghazi Plain, which stretches between Al Jabal Al Akhdar and the sea, and in the Jafara Plain in the north-west region. Although over-irrigation, some soils have led to increased soil salinization in these regions (Zurqani et al. 2019).

Libya's climate is dominantly influenced by both the Mediterranean Sea in the northern regions and the desert in the southern regions. In winter, the weather is cool with some rain on the coastal areas, and in the desert, the temperature may drop to sub-freezing at night. While there is almost no rainfall in the summer, as it is the major heat and drought period of the year. The Saharan area is very dry and hot in summer and cool and dry in winter. The average annual rainfall varies from region to region according to the geographic position and topography. Rainfall occurs during the autumn and winter months (October to March) (Ben Mahmoud 1995). The mean annual rainfall varies from zero millimeters in Southern Libya to 600 mm at Al Jabal Al Akhdar in the north-east (Yigini et al. 2013). The dominant features of rainfall are scantiness and variability in intensity from year to year (Fig. 1.2).

Water resources in Libya depend, to a large extent, on groundwater (about 97% of total water consumption). The sustainable development of the country is affected by the lack of renewable water resources as well as the overexploitation of fossil groundwater to meet the increasing demand for water. Other challenges include seawater intrusion, deteriorating water quality, and an inadequate institutional framework. These and other issues related to energy and food need more attention if the United Nations Sustainable Development Goals (SDGs) are to be achieved by 2030.

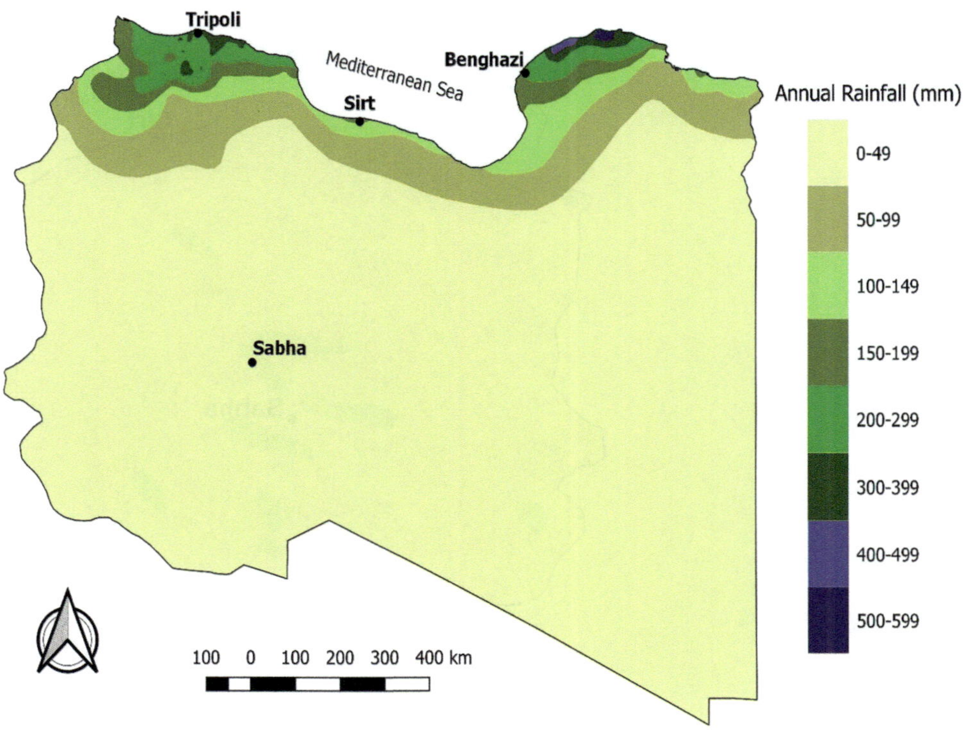

Fig. 1.2 Map of the annual rainfall in Libya adapted from (Yigini et al. 2013)

The most prominent natural attractions in the country are the Sahara Desert and the Mediterranean coast (General Water Authority 2014). In general, the topography of Libya slopes gently to the north. The mountains, which are located in the north-west, north-east, south, and central parts of the country, and are of low to medium latitudes (CEDARE 2014). The only hills are located near the border with Chad, where the Tibesti mountain massif rises to over 2,200 m. Only the coastal area, which accounts for 2% of the surface area, is suitable for agricultural activities. The fertile lands of the Jafara Plain in the north-west and Al Jabal Al Akhdar in the north-east account for more than 80% of food production from agriculture, which mostly depends on rainfall and groundwater. The remaining area consists of sand or gravel dunes, mountains, or marshes (salt marshes) with some scattered oases (Fig. 1.3).

The last census in 2006 estimated the population of Libya at 5.75 million (0.8 million non-Libyans) (LIB/00/004 2009; Zurqani et al. 2019). According to the United Nations World Population Prospects, this number increased to 6.87 million by June 2020 (World Population Review 2020). It is expected to reach 9 million people by 2030 (Environment General Authority 2008), which will increase the pressure on resources, especially food, water, and energy. The majority of the population lives along the Mediterranean coast, with a large proportion of them residing in major cities such as Tripoli and Benghazi.

1.2 Soil-Related Challenges in Libya and Potential Solutions

Libya is experiencing a severe problem of land degradation and desertification,

with one of the highest levels of deterioration in the world (43.7%) (Zurqani et al. 2019). Ongoing rapid urbanization along with the loss of fertile soil, the continued overexploitation of water resources, and overgrazing and the destruction of natural vegetation are the main causes of these environmental problems in the region.

1.2.1 Erosion

Soil erosion processes by both water and wind are one of the major soil environmental problems in the country and are responsible for land degradation. The harsh climatic conditions in Libya, including the shortage of land receiving sufficient rainfall beside a dry season of more than six months, have led to reduced vegetation, increased soil erosion, and widespread land degradation and made the region vulnerable to desertification. As a result, the region is traditionally classified as areas with high potential erosion rates. In spring and autumn, strong southerly winds that are locally known as (Gibli) blow from the desert, filling the air

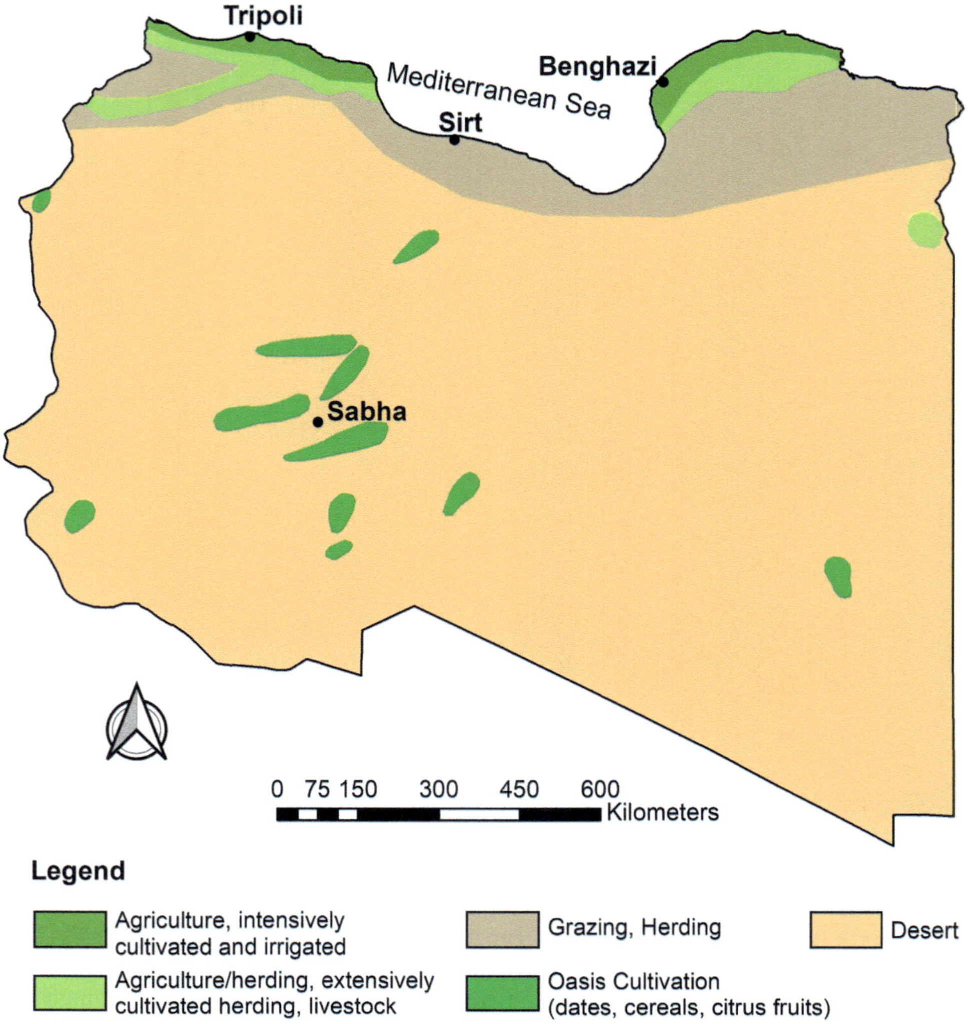

Fig. 1.3 Locations of agricultural activities in Libya (Adapted from: Zurqani et al. 2019)

with sand and dust and raising the temperature to about 50 °C in some areas. This wind activity is recognized as one of the major erosion factors in the county, which transports the eroded soils from one place to another (Fig. 1.4).

Soil erosion by wind takes place widely across all the country where most of the region is desert. The main factors contributing to soil erosion by wind hazard are: arid climate, the absence of adequate vegetation cover, soil texture, wind regimes, and the human factor (Yigini et al. 2013). Soil erosion by wind can remove several centimeters of the topsoil per year in the area that have sandy soils. Generally, soil erosion by wind occurs in the form of deflation within the littoral plain, while the soil erosion by water commonly occurs in the form of sheet washing, mainly within the Al Jabal Al Akhdar and on steeper slopes of Jabal Nafusah because most of the vegetation cover has been cleared by over-grazing and over-cultivation. The soils of Libya are subject to severe, moderate, and light of both water and soil erosion by wind as shown in Table 1.1a and b.

The degree of erosions in Libya shows that 592,053 ha and 859,634 ha have been affected by both soil erosion by water and wind, respectively, and heavy exposure to the erosions is shown in the western part of the country of both soil erosion by water and wind with total areas of 304,700 ha and 582,800, respectively, Table 1.1a and b. The degree of soil erosion by water is based on the precipitation characteristics (intensity and seasonality), the vegetation density, slope degree, and soil moisture. Figure 1.5 shows how severe soil erosion by water is a hazard in the mountainous region of Tarhuna in the north-western part of Libya. The lack of vegetation cover in the area cannot protect the soils from being affected by the precipitation intensities, which tend to produce passably large runoff volumes, leading to an increase the soil erosion.

The soil erosion by water and wind increase the exposure of the topsoil, causing the removal of fertile and loss of nutrient and seed stores, which eventually lead to the exposure of barren, locally hard-setting subsoils that resist

Fig. 1.4 Soil erosion by wind activity in Fiji, southern region of Libya (Photo courtesy: Abdalla Adda)

Table 1.1a **a** The affected areas of soil erosion by water in Libya (Environment General Authority 2002)

Regions	Degree of erosion by area (ha)			
	Light	Moderate	Severe	Total
Western	165,600	164,600	64,500	304,700
Middle	–	2,453	–	2,453
Eastern	241,700	41,500	1,700	284,900
Total	407,300	208,553	66,200	592,053

Table 1.1b **b** The affected areas of soil erosion by wind in Libya (Environment General Authority 2002)

Regions	Degree of erosion by area (ha)			
	Light	Moderate	Severe	Total
Western	180,000	266,400	136,400	582,800
Middle	160,838	49,494	4,402	214,734
Eastern	53,500	8,400	200	62,100
Total	394,338	324,294	141,002	859,634

revegetation. In this case, the rehabilitation of these soils becomes critical or impossible.

Over the past three decades, Libya has taken measures to control soil erosion (e.g., afforestation, sand dune stabilization with dry materials, petroleum emulsion, and synthetic rubber) (Yigini et al. 2013). Yet further steps are needed to help prevent or stop the erosions. Farming systems that rotate crops can effectively reduce soil erosion, sharply reducing the period of time that the soil is exposed to water and wind. Soil erosion by wind can be reduced by planting rows of trees at right angles that can prevail the wind to break its force and provide shelter for the field just beyond. Soil erosion by water can also be controlled by using mechanical measures, such as terracing, which is by far considered one of the most effective practices for controlling runoff and reducing soil erosion.

1.2.2 Salinity

Salinization is the primary type of desertification and soil degradation in Libya. Soil salinity problems started to be noticed in the country after the increasing agricultural activities, coupled with an increasing population (Zurqani

Fig. 1.5 Soil erosion by water hazard in the mountainous region of Tarhuna in the north-western part of Libya (Photo courtesy: Mohamed Alwasa)

et al. 2012). Most of the coastal areas are at risk because of the seawater intrusion resulting in salinity problems, caused by subsequent irrigation through local wells. Seawater intrusion has led to a high percentage of salts in some irrigation water, which in turn has led to the deterioration of fertility and low productivity of the soil. For instance, a critical problem has been observed in the north-western coastal strip areas, which is one of the most important agricultural lands in Libya that holds about 50% of the total irrigated land in the country and produces about 60% of the total agricultural production. Saline soils in Libya cover approximately 12% of the north region, 16.5% of the west region, and 23.4% of the middle region (Ben Mahmoud 1995). Many studies related to soil salinity assessment and mapping were conducted in Libya using different methods (Zurqani et al. 2012; Nwer et al. 2013, 2014; Atman and Lateh 2013; Zurqani et al. 2018). The results of these studies showed that most coastal areas are affected by salinity. Besides, the presence of closed depressions that are usually near to the water table and covered with a salt crust, known as Sabkhas (Fig. 1.6).

These Sabkhas are covered with a thin water layer during the winter season, but during the hot season, they are always dry, and their basins are usually covered by a salt film

Fig. 1.6 Saline soils (Sabkha) in the coastal area, Zuwarah, north-western region of Libya (Hamdi A. Zurqani 2009)

(Fig. 1.7). Zurqani et al. (2012) assessed the spatial and temporal variations of soil salinity in the north-western region of Libya using remote sensing and GIS and reported that the saline soils were decreased throughout the study period with about 0.45% per year, and these changes were explained due to the exposure of soil erosion by wind and sand encroachment over the study area.

There are numerous techniques that have been used to track and monitor these environmental disturbances. However, remote sensing holds the advantage of observing and

Fig. 1.7 Saline soils (Sabkha) with surface salt crust in Twewa, southern region of Libya (Photo courtesy: Abdalla Adda)

monitoring the earth surface's change because of the large spatial coverage, high time resolution, and wide availability. The newly available geospatial technology such as Google Earth Engine can be used to continuously monitor the environmental issues and provide management strategies that are able to react to and mitigate the environmental consequences of these challenges (Zurqani et al. 2019).

1.2.3 Pollution

Soil contamination or soil pollution as part of land degradation is caused by the accumulation of heavy metals near industrial zones, mining waste, disposal of high mineral waste, leaded gasoline and paints, land use for fertilizers, animal manure, sewage sludge, pesticides, and sewage disposal (Wuana and Okieimen 2011). The study by Mlitan et al. (2013) revealed that in the Zliten area around a cement factory, the soil physicochemical characteristics (pH, water content, and organic matter) indicated a strong influence by cement dust that has settled on the soil from the factory. Furthermore, Banana et al. (2017) reported that the Abu-Kammash area in the north-western region of Libya is contaminated with several toxic elements and petrochemical compounds, and the soil should be not be used for agriculture until it is remediated to ensure human health.

Since the inception of the Libyan conflict until now, a huge amount of gases, very fine particulates, inorganic/organic chemicals, and unknown toxic substances have continued to spread into the environment (Fig. 1.8). These compounds often contain high toxicities and carcinogenic properties and can expose humans who consume contaminated foods (Wang et al. 2013). There are many remediation techniques available for environmental restoration activities. For instance, Streche et al. (2018) suggested that using electrochemical treatment would be feasible and successful for soil remediation. In addition, the integration of sewage sludge into artificially polluted soils with crude oil can lead to increased biodegradation by 45–60% (Choroma et al. 2010).

1.2.4 Climate Change

Climate factors significantly contribute to both desertification and water scarcity, and global warming undoubtedly exacerbates these problems. Warmer air and water temperatures, and sea-level rise are expected to alter local ecosystems and agricultural productivity to the countries that have a coastline on the Mediterranean Sea. Libya has the longest Mediterranean coastline among African nations with about 1,770 km. Climate change will lead to severe weather events for Libya, affecting both the intensity and frequency of climatic factors such as temperature, precipitation, and droughts.

The increase in air and water temperatures will lead to a global sea-level rise. The United States National Oceanic and Atmospheric Administration scientists projected that sea

Fig. 1.8 Crude oil spillage near one of the Libyan oil fields in Jikharra, north-eastern region of Libya (Photo courtesy: Dr. Naji Shoaib)

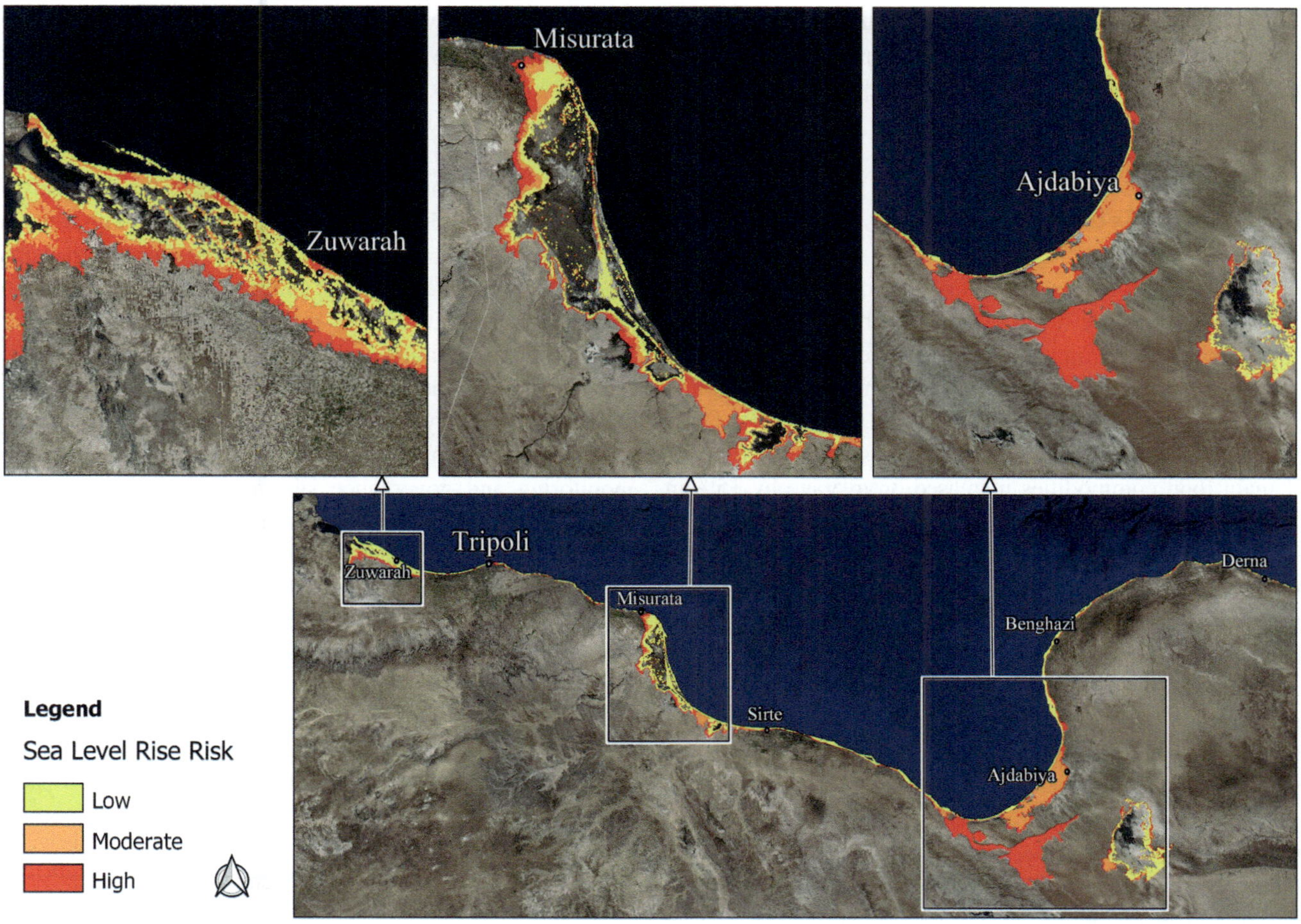

Fig. 1.9 The level of risk sea-level rise poses along Libyan coastlines areas

levels could rise by 0.5 m by 2050 (Sweet et al. 2017). Sea-level rise will erode shorelines, inundate low-lying coastal areas, and impair the operations of coastal infrastructure. Figure 1.9 shows the sea-level rise risk in coastal areas in Libya with projections of 1 m (low), 3 m (moderate), and 6 m (high), which will result in losses of land area equaling 700 km^2, 3200 km^2, and 7200 km^2, respectively. Cities near the sea are expected to be affected by sea-level rise of one meter, such as Zuwara, Misurata, Ras Lanuf, Aqilah, Brega, and Zaytiniyya. Sea-level rise will increase the proportion of damage in these cities (Zurqani et al. 2019).

Large-scale agricultural activities near the coastal areas, coupled with an increasing use of irrigation with the overdrawn supplies of fresh groundwater and the increase in the population, have caused widespread seawater intrusion problems. Many areas in Libya are currently vulnerable to land degradation and the situation will worsen if climate change and human activities continue to accelerate land degradation processes (Saad et al. 2011). For example, salinization in the Jafara Plain north-west region has reduced wheat yields from 5 tons ha^{-1} in the 1980s to just 0.5 tons ha^{-1} by 1987 (FAO 2015). Most farms in this region are irrigated by individual wells using electric pumps, and these farms have been severely affected by seawater intrusion into the aquifer, particularly orange farm operations. The increase of hot temperatures and evaporation processes will lead to an increase in the land areas affected by salinity in the country. It is necessitated to show that Libya is affected by climate change in many different ways, in particular, crop production and food security, pastoral lands and livestock production, water resources, etc. The frequent increase of sand storms can be explained by the recent climate change scenario, which includes declining precipitation and increasing temperature, followed by overgrazing and over-cultivation in marginal lands that cause severe soil erosion.

Climate change poses serious interlinked challenges in the country, where climate change and land degradation are closely related. Thus, the potential of conservation agriculture to minimize the impact of climate change needs thoroughgoing investigation. As a matter of fact, the impact of climate change on the soil system should be monitored on a regular basis. These data are needed not only as input for modeling and scenario analysis but also for monitoring, impact assessment, and characterizing the food production systems in target sites, while also viewing the productivity and economic losses in the face of the effects of climate change in the region.

1.2.5 Food Security

Healthy soils are the foundation of the food system and are necessary to produce healthy food to achieve food security in the country. Zurqani et al. (2019) reported that Libya imports approximately 80% of its food consumption requirement (e.g., wheat flour, corn oil, and milk). Quite recently, food prices continuously increased, where the cost of main food commodities increased on average by 15% in October 2018 when compared to the same month in 2017 (World Food Programme 2018). Furthermore, the Multi-Sector Needs Assessment (2018) reported that most households spend over 50% of their expenditure on food. This insecurity of food is driven by the lack of economic access, not food availability, and varies greatly among the administrative districts of the country. Where in the south-east region of the country, almost 70% of the households are food insecure. Food demand is expected to significantly increase in forthcoming decades due to accelerated population growth, expected to be approximately 1% annually (Zurqani et al. 2019). Libya must prioritize combatting soil degradation. Lately, massive amounts of fertile agricultural land in Libya are degraded because of soil erosion, salinity, nutrient depletion, pollution, and the other climate change-related extreme weather events. Out of the 2.1 million hectares of land suitable for agriculture, 1.8 million hectares are classified as arable and 300,000 ha under permanent crops, mostly fruit trees (FAO 2019).

The Libyan government should make this issue a priority by increasing investment in holistic approaches to building healthy soils. Yet several steps need to be done to ensure healthy soils and thus food security (e.g., support agricultural promotion policy strategies to reverse widespread soil degradation; creating a map of the health status of the country's soils where these data can be used to inform policies and develop recommendations for Libyan farmers about best practices and which fertilizers and seeds to use; etc.).

1.3 Conclusions

Soil is a finite resource, and its loss and degradation are not recoverable within a human lifespan. Without healthy soils, we would not be able to grow our food. This chapter provides a brief introduction about the soils of Libya and illustrates the soil-related challenges that currently the country is facing, which need to be addressed to improve the availability of services to support human needs. Erosions, salinization, land degradation, and climate change are the largest challenges that currently are facing soils and future agricultural development in Libya. The new geospatial technology such as Google Earth Engine can be used to continuously monitor these environmental disturbances to help provide the best possible management strategies that might be able to react to and mitigate the consequences of these disturbances.

The harsh climatic conditions and poor soils limit agricultural output, where domestic food production meets only about 25% of Libyans' demand. The country has recently underscored the importance of implementing a food security monitoring and governance system. To do this, the government should put in place legal frameworks to regulate the use of soil resources. In particular, farmers should be educated about and encouraged to adopt these proposed soil health-building practices. Also, the continuous support at the country level from both public and private institutions is essential to make these sustainable practices in the long term.

Acknowledgements The author acknowledges Dr. Naji Shoaib, Abdalla Adda, and Mohamed Alwasa for providing selected photos with their permissions to use them in this chapter.

References

Banana AA, Al-Gheethib AA, Nomanc EA, RMSR M (2017) Assessment of soil pollution by toxic metals and petrochemical compounds in western Libya. Desalination Water Treat 83, 272–276. https://doi.org/10.5004/dwt.2017.20781, https://www.deswater.com/DWT_abstracts/vol_83/83_2017_272.pdf

Ben Mahmoud KR (1995) Libyan soils (their genesis, classification, properties, and agriculture potentiality), 1st edn. National Research Scientific Organization, Tripoli Libya (in Arabic)

CEDARE (2014) Libya Water Sector M&E Rapid Assessment Report. Monitoring and Evaluation for Water in North Africa (MEWINA) project, Water Resources Management Program, CEDARE. https://www.humanitarianresponse.info/sites/www.humanitarianresponse.info/files/assessments/libya_water_sector_me_rapid_assessment_2014.pdf. Accessed 2 Sept 2020

Choroma M, Hosseini SS, Motamedi H (2010) Bioremediation of crude oil polluted soil as affected by sewage-sludge. Paper presented at: 19th World Congress of Soil Science, Soil Solutions for a Changing World, Brisbane, Australia. https://iuss.org/19th%20WCSS/Symposium/pdf/0691.pdf

Environment General Authority (EGA) (2002) The first environmental report. Tripoli, Libya

Environment General Authority (EGA) (2008) National Strategy for Sustainable Development report. Tripoli, Libya

FAO, GIEWS—Global Information and Early Warning System, Country Briefs, Libya. 2019 Feb [cited 2019 Mar 12]. https://www.fao.org/giews/countrybrief/country.jsp?code=LBY

FAO (2015) Status of the World's Soil Resources (SWSR)–Main Report. Food and Agriculture Organization of the United Nations and Intergovernmental Technical Panel on Soils, Rome, Italy, 650. https://www.fao.org/3/a-i5199e.pdf

General Water Authority (2014) Water and Energy for Life in Libya (WELL), Project funded by the European Commission No. 295143, FP7, Libya. https://cordis.europa.eu/project/id/295143/reporting. Accessed 2 Sept 2020

LIB/00/004 (2009) Atlas of Natural Resources for Agricultural use in Libya. https://www.fao.org/geospatial/resources/detail/en/c/1024739/. Accessed 2 Sept 2020

Mlitan AB, Alajtal AI, Alsadawy AM (2013) Toxicity of heavy metals and microbial analysis of soil samples collected from the area around Zliten cement factory. Open J Air Pollut 2(01):25. https://www.scirp.org/journal/PaperInformation.aspx?PaperID=29015

Nwer B, Ali A, Zurqani H (2013) Soil salinity mapping model developed using RS and GIS in Libya. In: Global science & technology forum. https://www.geosp.net/?research=soil-salinity-mapping-model-developed-using-rs-and-gis-in-libya

Nwer B, Zurqani H, Rhoma E (2014) The use of remote sensing and geographic information system for soil salinity monitoring in Libya. GSTF J Geol Sci 1:1. https://doi.org/10.5176/2335-6774_1.1.5

Saad AMA, Shariff NM, Gairola S (2011) Nature and causes of land degradation and desertification in Libya: need for sustainable land management. Afr J Biotechnol 10(63): 13680–13687. https://www.ajol.info/index.php/ajb/article/view/96799

Streche C, Cocârţă DM, Istrate IA, Badea AA (2018) Decontamination of petroleum-contaminated soils using the electrochemical technique: remediation degree and energy consumption. Sci Rep 8 (1):3272

Sweet WV, Kopp RE, Weaver CP, Obeysekera J, Horton RM, Thieler ER, Zervas C (2017) Global and regional sea level rise scenarios for the United States. https://ntrs.nasa.gov/search.jsp?R=20180001857

Wang XT, Miao Y, Zhang Y, Li YC, Wu MH, Yu G (2013) Polycyclic aromatic hydrocarbons (PAHs) in urban soils of the megacity Shanghai: occurrence, source apportionment and potential human health risk. Sci Total Environ 447:80–89. https://doi.org/10.1016/j.scitotenv.2012.12.086

World Food Programme (WFP), United Nations (UN), Libya - Market Analysis Report, October 2018. https://www.wfp.org/content/libya-market-analysis-report-october-2018. Accessed 13 March 2019

World Population Review (2020) Libya population live. https://worldpopulationreview.com/countries/libya-population. Accessed 2 Sept 2020

Wuana RA, Okieimen FE (2011) Heavy metals in contaminated soils: a review of sources, chemistry, risks and best available strategies for remediation. ISRN Ecol. https://doi.org/10.5402/2011/402647

Yigini Y, Panagos P, Montanarella L (2013) Soil resources of Mediterranean and Caucasus countries. Office for Official Publications of the European Communities, Luxembourg, p 237. https://publications.jrc.ec.europa.eu/repository/bitstream/JRC80174/lb-na-25-988-en-n.pdf

Zurqani HA, Mikhailova EA, Post CJ, Schlautman MA, Elhawej AR (2019) A review of Libyan soil databases for use within an ecosystem services framework. Land 8(5):82

Zurqani HA, Mikhailova EA, Post CJ, Schlautman MA, Sharp JL (2018) Predicting the classes and distribution of salt-affected soils in Northwest Libya. Commun Soil Sci Plant Anal 49(6):689–700. https://doi.org/10.1080/00103624.2018.1432637

Zurqani HA, Nwer BA, Rhoma AR (2012) Assessment of spatial and temporal variations of soil salinity using remote sensing and geographic information system in Libya. In: Conference proceedings on 1st annual international conference on geological & earth sciences. https://doi.org/10.5176/22513361_geos12.64 https://dl4.globalstf.org/wpsc-product=assessment-of-spatial-andtemporal-variations-of-soil-salinity-usingremote-sensing-and-geographic-informationsystem-in-libya

Dr. Hamdi A. Zurqani is one of the faculty members at the Department of Soil and Water Sciences, Faculty of Agriculture, University of Tripoli, Tripoli, Libya. Dr. Zurqani is a recognized expert as a result of his internationally acclaimed work in the areas of environmental information science, remote sensing, land evaluation, sustainability, pedology, and soil science education. He has conducted research across the world, including the United States of America and Africa. Dr. Zurqani is a distinguished soil scientist with a wide range of scientific and working experiences in Libya and abroad. He received his M.Sc. (2010) from the University of Tripoli, Tripoli, Libya and Ph.D. (2019) from Clemson University, Clemson, SC, USA. His major research and teaching activities at the University of Tripoli have focused mainly on soil genesis and classification and the environmental information sciences (Remote Sensing and Geographic Information System). He has published broadly in many journals (e.g., Nature "Scientific Reports", Geoderma; International Journal of Applied Earth Observation and Geoinformation; Journal of Geological Sciences; Land; Frontiers in Environmental Science; Communications in Soil Science and Plant Analysis; and others). Dr. Zurqani is a member of the Editorial Board for Remote Sensing (MDPI) Journal, counseling outcome and research evaluation. He also was appointed to serve as a Guest Editor for the Special Issue "Applications of Remote Sensing in Earth Observation and Geo-Information Science". In addition, Dr. Zurqani conducted peer-review for many journals including Journal of Environmental Informatics, Applied Sciences, SN Applied Sciences, Remote Sensing, Heliyon, Geosciences, Land, Water, Agronomy, Agriculture, Sustainability, Arid Land Research and Management, International Journal of Environmental Research and Public Health, Natural Hazards, and Conference of the Arabian Journal of Geosciences. He is also one of the authors of the lab manual entitled, "GIS Exercises for Natural Resource Management". Dr. Zurqani has been the recipient of numerous awards and honors: Recipient of Douglas R. Phillips Award for Graduate Students, Department of Forestry and Environmental Conservation, Clemson University, April 12, 2019; the First Place Best Judged Poster (CAFLS) at the GRADS 2019: Clemson Student Research Forum on April 4, 2019; the Second Place Poster at the 11th Clemson Biological Sciences Annual Student Symposium, April 6, 2019; the Second Place Best Judged Poster at the Clemson Student Research Forum on April 4, 2018; and the Third Place Poster at the 9th Clemson Biological Sciences Annual Student Symposium, February 25, 2017. Dr. Zurqani conducts cutting-edge research in the field of environmental information science, remote sensing, land use management/planning, change detection of landscape degradation, and geographic information system (GIS) models. He has focused on his research efforts on the development of new technologies in the field of environmental information sciences, geo-intelligence (advanced geo-information science and earth observation, machine and deep learning, and big data analytics), remote sensing, land evaluation, pedology, land use management/ planning, monitoring and evaluating sustainable land management, change detection of landscape degradation, and geographic information system models.

2 The History of Soil Mapping and Research

Mukhtar M. Elaalem, Hamdi A. Zurqani, and Khaled R. Ben Mahmoud

Abstract

In this chapter, the authors outline the development of historical soil knowledge, education, and research in Libya. In addition, they reviewed the soil institutes and universities conducting research on soil science. This review has shown that Libya's capacity to absorb new technologies is now greater than before. Due to the continuous rise in educational levels, Libya is more integrated than ever in advanced science and technical education systems, competing with more advanced countries in modern technologies. Soil resource studies and mapping in Libya which have been carried out in the last decades have been discussed in detail. Extensive soil studies were given mainly to areas with high and medium potential for agriculture in the northern part of the Libyan Coastal zone and to only small, scattered areas in the southern desert. Research projects in Libya and achievements in the field of Soil Mapping are reviewed at the end of this chapter. Soil taxonomic units of Libya and their major properties are found in Chap. 3.

Keywords

Soil survey • Soil education • Soil mapping • Soil research • Pedology • Libya • North Africa

2.1 Introduction

When talking about the history of the development of knowledge in the field of agriculture in Libya, especially in one of its important branches in the field of soil resources, it must be noted that Libya was an Italian colony for 32 years and did not become independent as a state ruled by the Libyans until 1952. That is why there were no serious attempts to develop agricultural knowledge and establish specialized institutions or Libyan cadres before the independence period. This does not mean that we ignore the traditional knowledge that the local people have accumulated from farmers and shepherds throughout the distant history or the experience of Italian colonialism (1911–1945) in the agricultural development in the north-west and north-east regions during the Italian agricultural settlement. Given the extreme poverty that Libya suffered from after independence in the 1950s, the Libyan government could not do much in the area of developing the agricultural sector in a country that lacks specialized institutions and scientific cadres, and so heavily relied on foreign aids from a number of governments and international organizations.

The young Libyan government began establishing a ministry of agriculture and some medium agricultural institutes as well as sending a limited number of scholarships for university studies abroad. The financial situation continued to the mid of 1960s, that is, after the discovery of oil in Libya and the beginning of exports, allowed the Libyan economy to improve. Then, Libya began to make great strides in all areas of economic and social development, and agriculture gained ample benefits from that effort. The Faculty of Agriculture was established at the University of Tripoli in 1966 and gradually began to develop in the field of studies and scientific research in all fields of agriculture, especially soil resources. In the beginning, it was done with the help of non-Libyan experts in all institutions of the Ministry of Agriculture and Faculties of Agriculture, and then beginning in the 1970s, contracts were made with some

M. M. Elaalem · H. A. Zurqani (✉) · K. R. Ben Mahmoud
Department of Soil and Water Sciences, Faculty of Agriculture, University of Tripoli, Tripoli, 13538, Libya
e-mail: h.zurqani@uot.edu.ly

M. M. Elaalem
e-mail: M.Elaalem@uot.edu.ly

K. R. Ben Mahmoud
e-mail: benmahmoud.kh@gmail.com

H. A. Zurqani (ed.), *The Soils of Libya*, World Soils Book Series,
https://doi.org/10.1007/978-3-030-66368-1_2

foreign companies and consulting offices to carry out inventory and mapping studies of soil and lands, and from the 1980s until now, Libyan experts have taken over the development of institutions working in the field of soil resources. In this chapter, we review the historical development of soil studies, mapping and scientific research in Libya.

2.2 History of Soil Knowledge, Education, and Research

2.2.1 Ancient History of Soil Knowledge

Over the past years, old Libyans have accumulated considerable knowledge and experience in the selection of land suitable for rainfed agriculture, in the organization of grazing craft, in water harvesting techniques, in the search for groundwater presence, hand-dug wells, how to extract water from these wells and in choosing the appropriate irrigation system, and in the selection of crops to suit all the different regions in Libya, each according to the prevailing environment. This knowledge and experience were acquired through trial and error and through the transfer of experiences from other people who colonized and settled in Libya throughout history. Here are some of these experiences:

a. Grazing in the life of the Libyan man over many centuries was not only a profession but also a nature of life and a major source of livelihood. Pastoralist communities, despite their illiteracy, realized that grazing was not only a process of grazing by livestock but also a kind of coexistence and mutual benefit between humans, animals, plants, and land. The best evidence of pastoral communities' keenness on providing food for animals was their success in producing breeds of livestock with good characteristics that can be formed only in the conditions of availability of pastures. These strains remain a challenge to this day. The inhabitants of the Bawadi instinctively concluded that the availability of pastures led to the prosperity of cattle, their livelihoods, and glory. This instinct led these communities to develop a system of **Al Hamah*** (an old Bedouin herding system in which pastoral lands are divided annually into lands where grazing is allowed and the other is left without grazing to give an opportunity to develop pastoral plants for the coming year, and so on) (Fig. 2.1). Alhamah system was the first principle of rangeland management and a sound basis for the improvement and development of pasture resources and soil and water conservation in the world.

b. In rainfed agriculture in arid regions of Libya with rainfall of less than 200 mm per year, farming communities select agricultural lands in relatively low areas and in the bottom of dry valleys that receive runoff from neighboring areas and low rainfall. This experience was acquired through trial and error (Fig. 2.2).

c. Over the past years, human communities living in the Northern mountains of Libya; (i.e., western region and Al Jabal Al Akhdar) have gained experience in the distribution of surface water and the use of water harvesting techniques in primitive ways to conserve water in the soil or to collect it for drinking or supplementary irrigation (Fig. 2.3a and b).

Fig. 2.1 Old Bedouin Herding system

Fig. 2.2 Bottom of dry valley

Fig. 2.3 Collection water: **a** Drinking cattle, **b** Drinking people

d. In irrigated agriculture, old Libyans were able to dig shallow underground wells by manual methods in the northern and southern oases. They used animals such as donkeys and camels to extract water. Also, old Libyan farmers found a suitable way to irrigate their vegetables and fruit trees that are suitable in dry and hot climates, which is within the scope of flood irrigation by dividing the land into tables (Figs. 2.4, and 2.5).

2.2.2 The Experience of the Italian Agricultural Settlement in Libya

Between 1911 and 1943, Libya was occupied by Italy, and in the 1920s, the military ruler of Libya, de Bono (1928) and then Marshal Balbo (1930) launched the Agricultural Settlement Program in Libya which was based on the results of some reconnaissance and intensive detailed

Fig. 2.4 Traditional wells

surveys. Some of these studies may be found in the Italian government libraries. The agricultural survey was done for the territory of Tripolitania and Cyrenaica, including groundwater studies and land suitability for cultivation in these two regions.

In these two regions, the processing of the farms settled no more than 455 Italian families brought from Italy, a total of about 1,780 farmers in the regions of Tripolitania. The number tripled by the end of 1933 to 1,500 families, or 7,000 farmers. For the region of Cyrenaica, the situation was completely different; with only 429 Italian farmers there, due to the safety situation in the 1920s and early 1930s. They planted olive trees, fruit trees, and cultivated tobacco, especially in the Coastal areas such as Bianki, Zahra, Tripoli, Al Gharaboli, Al Khumis, Gharian, Tarhona. In other areas where water and good soils are available, they planted olive trees, grapes. almonds and many fruit trees surrounded by wind brake trees, which still stand in these areas.

In this experiment, the Libyans were briefed for the first time on modern farming methods, from dividing the farm to a rainfed section and irrigated section. In the rainfed section (75% of the area of the farm), olive or other trees were planted at large distances, which were devoted to the cultivation of wheat and barley crops (Fig. 2.6).

In the irrigated section, deep wells were dug for the first time, providing the infrastructure of the irrigation systems that were followed in that period. These later switched from flood irrigation to sprinkler irrigation and in each farm, attention was also paid to animal production; especially cattle breeding for meat and milk production. This area served as

Fig. 2.5 Old irrigated farm

Fig. 2.6 Olive trees integrated with wheat or barley (Rainfed System)

extension fields and indicators for agricultural development for the Libyan Ministry of Agriculture technicians and post-independence Libyan farmers (i.e., the 1950s and 1960s) (Farler 2006).

2.2.3 Development of Education in the Field of Soil Sciences and Resources

By the end of the 1950s, the Libyan engineers began to hold Bachelor's degrees in agricultural sciences in the fields of plant protection, crops, horticulture, agricultural economics, but in soil science, the first batch of Libyan scientists to obtain their Bachelor's degrees in the field of soil was in 1969, namely Khaled Ben Mahmoud, Abualqasim Albaruniu, Abdeljalil Youssef, from Cairo University, and Sadiq Ben Yazid from Baghdad University (Fig. 2.7a–d).

In 1966, the Faculty of Agriculture at the University of Libya in Tripoli was launched. Initially, it contained four main departments, one of which was the department of Soil and Water, which was founded in 1967. The Soil and Water department started to teach a number of courses, which

a) b) c) d)

Fig. 2.7 **a**. Prof. Khaled Ben Mahmoud, **b**. Prof. Abualqasim Albaruniu, **c**. Prof. Abdeljalil Youssef, **d**. Prof. Sadiq Ben Yazid

include introduction of soil science, soil physics, soil chemistry, soil genesis and classification, soil fertility and plant nutrition, soil microbiology, soil and water conservation, and irrigation science and groundwater hydrology. It should also be noted that all the faculty Staff members in the soil and water department were of non-Libyan nationalities (American, Indian, Iraqi and Egyptian). In addition to this, the first batch of graduates in the department of soil and water sciences consisted of 10 graduates in 1970.

Later, other colleges and higher institutes for agriculture were established in different regions of Libya, which included all the various agricultural fields, including a department of soil and water. To date, the number of colleges and higher institutes of agriculture has reached 10.

These colleges and institutes graduate hundreds of graduates specializing in soil and water and working in all relevant state institutions as well as the private sector. Table 2.1 shows institutes and faculties conducting research on soil and water science.

At the end of the 1980s, the Soil and Water department in Tripoli was separated into two divisions, namely soil and water branches. A number of courses were added, which is, Libyan soils, Soil fertility, Plant nutrition, Reclamation of Saline soils, and Remote Sensing and GIS.

In addition, the Ministry of Agriculture dispatched several university graduates to international universities, especially the USA, to obtain master's and doctoral degrees in agricultural sciences in the late 1950s and early 1960s. Also, Prof. KhairyAl Saghir was the first graduate student to arrive in Libya with a Ph.D. in Crop Science (Fig. 2.8a).

At the end of the 1960s, the Libyan University in Tripoli and Benghazi worked to send a large number of graduates from the faculties of science and agriculture to obtain doctorate degrees in various agricultural fields and later become staff members in the faculties of agriculture.

It should also be noted that Prof. Gilani Abdelgawad was the first graduate student to hold a Ph.D. level in soil chemistry in 1973 (Fig. 2.8b), and Prof. Khaled Ben Mahmoud was the first graduate student to obtain a Ph.D. in 1977 in the field of soil genesis, classification, and mapping (Fig. 2.8 c).

The first postgraduate degree program in the faculties of agriculture in Libya in the department of soil and water in the faculty of agriculture at University of Tripoli began in 1980. The first master's degree was awarded locally in soil chemistry in 1984 to Dr.MahjoubAl Qubi. Then, the program was developed and expanded in the majority of the faculties of agriculture in Libya. However, the program of obtaining a Ph.D. is still stalled in Libyan universities, and the Libyan government is still giving scholarships to Libyan students to apply for MSc and Ph.D. programs in the areas of soil science. These scholarships may take place in Libya or in overseas universities [**Authors collection**].

Table 2.1 Institutes and faculties conducting research on soil and water

Institute/faculty	Year of establishment
Higher and Intermediate Institute of Agricultural Technology, Ghairan, Tripoli	1953
University of Tripoli, Faculty of Agricultural	1966
University of Omar Al Mukhtar, Faculty of Agricultural	1975
University of Sebha, Faculty of Agricultural	1975
University of Sirte, Faculty of Agricultural	1994
University of Zawia, Faculty of Veterinary Medicine and Agricultural Sciences	2000
University of Alzeetona, Faculty of Agricultural	2001
University of Bani Walid, Faculty of Agricultural	2000/2001
University of Benghazi, Faculty of Agricultural, Solouq Branch	2010
University of Misurata, Faculty of Agricultural	2015
University of Misurata, Faculty of Agricultural	2015

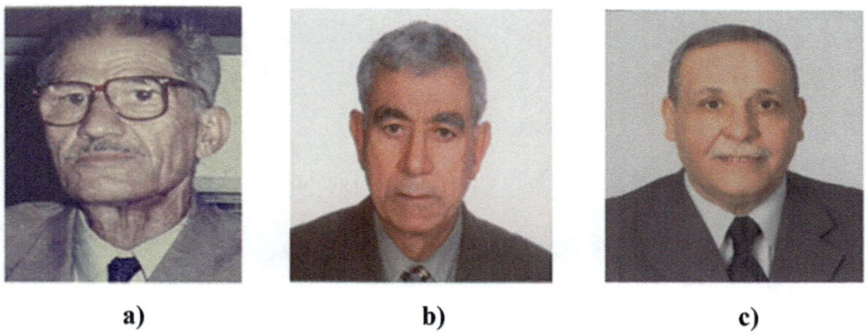

Fig. 2.8 **a** Prof. KhairyAl Saghir, **b** Prof. Gilani Abdelgawad, and **c** Prof. Khaled Ben Mahmoud

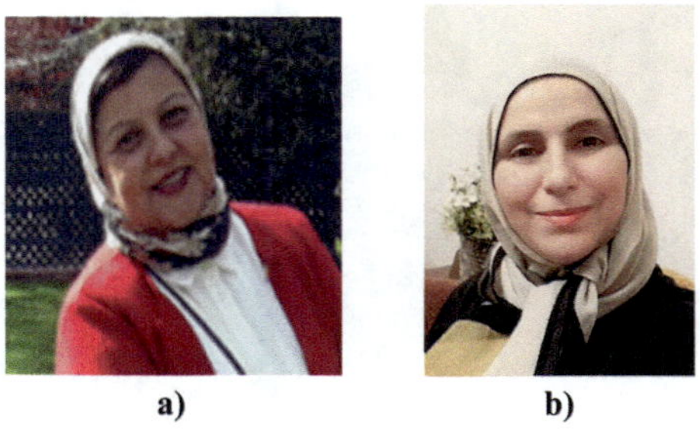

Fig. 2.9 **a** Dr. Eman Ali Ferjani, **b** Mrs. Jamila Suleiman Elarabi

At the beginning of 1972, Mona Khashba was the first female student to graduate with a bachelor's degree in Soil Sciences from the Department of Soil and Water Sciences, Faculty of Agriculture, University of Tripoli. Then, the number of females choosing to study soil and water sciences increased, reaching 423 in 2019 [**Authors collection**].

In 1996, Eman Ali Ferjani (Fig. 2.9a) was the first female student to graduate with a master's degree in Soil Microbiology from the Department of Soil and Water Sciences, Faculty of Agriculture, University of Tripoli. Meanwhile, Jamila Suleiman Elarabi (Fig. 2.9b) was the first female student to graduate with a master's degree in Soil Pedology in 2003. **[Authors collection]**.

So far, the number of females who have graduated from the Department of Soil and Water Sciences, Faculty of Agriculture, University of Tripoli, with a master's degree in different fields of soil and water sciences, is about 40, and only 8 of them are in Soil pedagogy. **[Authors collection]**.

The reason for the small number of female students (compared to males) enrolled in the Department of Soil and Water Sciences is primarily due to social reasons. What will students do after graduation? Of course, female as well as male graduates with a degree in soil sciences will work in the field and this is not desirable for Libyan families. Lately, after the development of Libyan society, and the nature of work in soil and water sciences had changed (laboratories, offices, Remote Sensing, GIS, etc.), the number of female students enrolled in the Department of Soil and Water Sciences has been noticeably increased. However, there is no significant increase in the number of female students in the Soil Pedology major, as students in this area are still required to do their work fundamentally in the fields.

2.2.4 Development of State Institutions Working in the Field of Soil and Land Mapping

In the early 1950s, the Ministry of Agriculture established a Soil and Water Conservation Authority. The main aims of this Authority are fixation of sand dunes, afforestation, and establishment of industrial forests, as well as the implementation of some methods of water harvesting and distribution of surface water in the Northern mountain areas (western Mountain and Al Jabel Al Akhdar).

In the 1960s, the Department of Soil and Water Conservation at the Ministry of Agriculture was formed on the ruins of the Soil Department. This department began work in the field of soil and land classification mapping. In this period of time, soil studies have not received much attention except for two studies [Authors collection].

At the beginning of the 1970s, agricultural development work in all regions of Libya began to accelerate with the establishment of the Council for Agricultural Development. Meanwhile, the soil department of the Ministry of Agriculture lacked specialists in the field of soil science, so the Council was contracted with international companies to carry out soil and water studies for different agricultural development areas in Libya. At the same time, foreign soil engineers were recruited in the department who carried out some soil studies and mapping in various other areas (will be discussed later).

In 1977, the Ministry of Dams and Water Resources was launched, and in 1980, a General Directorate of Soil department at the Ministry of Agriculture was created as an alternative to the soil department of the Ministry of Agriculture.

Fig. 2.10 Libyan and FAO experts

The General Directorate of Soil department specialized in conducting scientific studies and researches on soil, to improve its properties, reclaiming new lands for agriculture and protecting the soil from erosion, pollution and stress in order to obtain the best possible production of various agricultural crops. The General Directorate of Soil was also tasked with the technical development in the field of irrigation and drainage management, as well as preparing specialized technical people in the field of soil science. The soil department was renamed the Soil Management and Water Use in the 1980s (Ben Mahmoud et al. 1984).

In 2001, a natural resource mapping project for agricultural use and planning was established in Libya in collaboration with the FAO experts under the Ministry of Agriculture. The National coordinator of the project was Dr. Khaled R. Ben Mahmoud with a stirring committee consists of the following stake holders (Fig. 2.10).

- Ministry of Agriculture
- Department of Soil and Water Sciences, Faculty of Agriculture, University of Tripoli
- Agricultural Research Centre.
- Department of Soil, Water Authority.
- Surveying Department
- Department of Climate in Libyan National Meteorological Center
- Libyan centre for Remote Sensing and Space Sciences

.

The Natural Resource Mapping Project for Agricultural Use and Planning is specialized for (LIB00/004 2007):

- Digitizing all topographic, soil and land maps and consolidating the data and information they contain.
- Building databases of natural resources (soil, water, climate) and their management systems.
- Building a Land Resource Information Management System in Libya (LRIMS).
- Mapping land/vegetation cover in Libya.
- Monitoring and assessing land degradation, and then preparing maps and identifying areas affected by desertification that require rapid intervention.
- Preparing an appropriate national methodology for drought control and then reducing its effects.
- Developing different scenarios of early warning for food security risks under climate changes.
- Training some of local experts in the areas of GIS, remote sensing, land/vegetation covers and preparing digital maps.

Fig. 2.11 Dr. Mohammed Zuhni

2.2.5 Development of Soil Research in Libya

Agricultural research was established on small scale in the early 1940s at Sidi Al Masri near Tripoli to serve the Italian Agricultural Settlers. In the early 1950s to late 1960s, the Ministry of Agriculture became affiliated with the Doctorate of Plant and Animal Production, along with AR. In 1970, the Agricultural Research Centre (ARC) was established at Tripoli to serve as an umbrella organization for implementation of AR in the country with an adopted national AR work plan (the first General Director of ARC was Dr. Mohammed Zuhni, Fig. 2.11).

The Centre specializes now in scientific research in the field of agriculture and works to strengthen its assets and create the necessary technical elements, to be an effective tool for the exploitation of natural resources and protection and development of existing agriculture into modern agriculture. The National Agricultural Research (NARS) of Libya currently includes (El-azzabi 1999):

- Tree sets of scientific institutes namely are: Agricultural Research Center (ARS), Animal Studies and Research Center (ASRC), and Marine Biology Research Center (MBRC).
- Faculties of Agriculture and Veterinary Madison which are also involved in agriculture research.
- There are also a few other institutes involved in agricultural research such as:
- Nuclear Research Center (NRC).
- Libyan Center for Remote Sensing and Space Sciences (LCRSSS).
- The Center of Biotechnology (CBT).
- The Economic Studies Research Center (ESRC).
- National Genetic Research Bank (NGRB).

Soil research, in particular, was carried within the ARC by the Soil and Irrigation section. This section concentrated its efforts in areas of soil sciences and water uses other than soil genesis, classification, survey, and mapping. Soil survey and mapping were the responsibility of the Soil Department in the Ministry of Agriculture in cooperation with the staff members who specialized in this area at the universities. The research in the area of soil genesis and classification was mainly conducted by the staff members and their graduate students [**Authors collection**]. These research projects and achievements will be discussed in detail in the coming section of this chapter.

2.3 Soil Resources Studies and Mapping in Libya

According to a review of the available data on soil resources studies in Libya, several soil studies have been done, and these studies are shown below (Fig. 2.12).

2.3.1 Soil Resources Studies (Pre-1945s)

Soil studies that were carried out during the Italian colonial period for agricultural settlement did not yield much benefit due to language and poor storage (Ben Mahmoudet al. 1984).

2.3.2 Soil Resources Studies (1945–1970)

In this period of time, about 15 soil studies were conducted in Libya. One of the most important of these soil resources studies was conducted by Stewart in 1960, and this study covered the north-western region of Libya, which was known as Tripolitania (Stewart 1960). Stewart's study aimed to produce land capability maps at a scale of 1:100,000. In 1964, soil experts from FAO studied the soil and land in the region of Cyrenaica (Habert 1964).

2.3.3 Soil Resource Studies (1970–present)

During this period, about 300 soil resources studies were conducted, and these studies have covered large areas of Libya. The majority of these soil studies took place in the 1970s and early 1980s and most were done during the implementation of Libya's accelerated agricultural development. These studies were aimed to drive pedological and interpretive soil maps at different scales. As a result of these

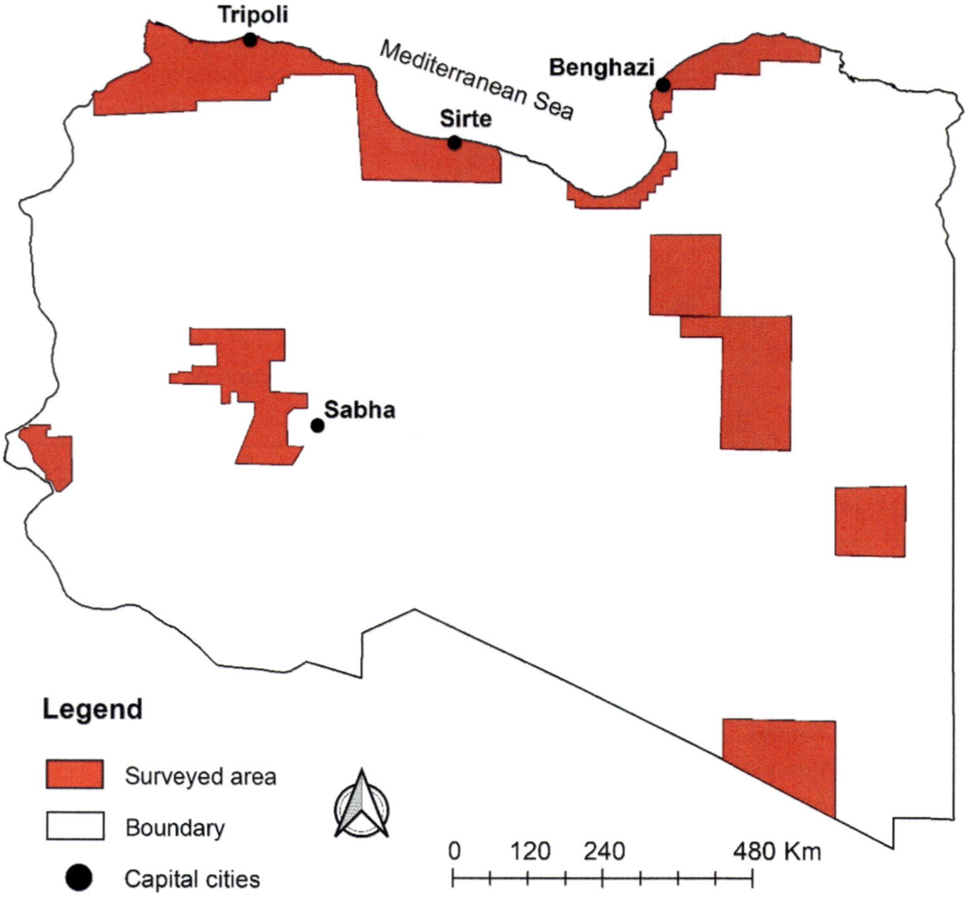

Fig. 2.12 Soil Resources Studies Surveyed in Libya (Adapted from, Ben Mahmoud 2001)

studies, the quality and productivity of land were determined, and then a number of agricultural projects were launched in Libya. It should be mentioned that these studies have been conducted by a number of companies of different nationalities. These soil studies are included below:

2.3.3.1 Studies of Soil Resources by French GEFLI Company (1971–1978)

These studies were implemented in 20 different locations in Libya and the mapping of soil surveys were exploratory, semidetailed and detailed. These studies considered the interpretation of aerial images and annual rainfall rates. GEFLI soil studies did not take into account the morphological characteristics of the soils and thus determinate their soil taxonomic units, but they conducted a survey to determine the quality of vegetation and the status of agricultural exploitation. Also, GEFLI studies used the French system for determining soil properties (GEFLI 1971–1978).

2.3.3.2 A Study of Soil Resources by Egyptian REGWA Company (1993 and 2005)

In 1993, REGWA company studied the soil and water resources, which are suitable in agriculture in the southern region of Libya (Al Gatroun, Tjerhi, and Al Magdoud). One of the main objectives of this study was the settlement of nomads in the settlement projects in the mentioned areas. This study included a derivation soil classification map, land capability map, and salinity and soil depth maps using the U. S. Soil Taxonomy system (REGWA 1993). In 2005, REGWA company also conducted a soil and water resources study in the Sara area with the same goals which have carried out in previous study (REGWA 2005).

2.3.3.3 A Study of Soil Resources by Russian Selkhozprom Export Company (1980)

This study was conducted in cooperation with the Libyan Ministry of Agriculture. This study was concerned with the study of coastal areas that receive rainfall rates of more than 200 mm per year (3 million hectares) according to modern soil classification of Russia.

This study was aimed at producing pedological and interpretive soil maps such as: soil classification, land capability, optimal land use, soil salinity, soil erosion, soil depth, and land suitability maps at scale 1:50,000. The Selkhozprom Export study also contained the results of field and laboratory studies, special technical reports and combined maps of western and eastern regions at a scale of

1:200,000. Meanwhile, this study examined the vegetation cover of 0.5 million hectares in the north-central region (Ajdabiya region) in order to determine the status of natural pastures (Solkhozprom Export 1980).

2.3.3.4 A Study of Soil Resources by Moroccan Telekart Company (1989)

This study was conducted in some areas of the Jafara Plain of Libya, such as Al Joush, Teji and Badr. This study was focused on vegetation cover for the purpose of developing pastures and also carried out some soil studies in the mentioned areas. The U.S. Soil Taxonomy system was used in producing soil classification maps and interpretive soil maps at a scale 1:50,000 (Tele Kart 1994).

2.3.3.5 Studies of Soil Resources of Candidate Areas for Agricultural Investment (1987–2007)

The studies were carried out by the soil team of the Man-Made River Water Investment Authority, in cooperation with a number of soil mapping experts at the University of Tripoli, headed by Dr. Khaled ben Mahmoud, Dr. Azzeddin R. Elhawej and Dr. Khalil Suleiman. There were 25 sites scattered in the north-west, north-central and north-eastern regions. These studies were conducted in the paths of the Man-Made River project and considered the criteria of the Modern US soil taxonomy. Through these studies, pedological soil maps and interpretive soil maps were obtained. These studies contain reports and maps at a scale of 1: 50,000 and 1:20,000 (Soil Survey Team, Man-Made River Investment Authority (Phase I and II, 1986–2007).

2.3.3.6 Studies of Soil Resources for the Agricultural Mapping Project in the north-western coastal areas, which has not covered by the Study of Russian Selkhozprom Export Company (National Consultative Office in corporation with in cooperation with a number of soil mapping experts at the University of Tripoli, headed by Dr. Khalid ben Mahmoud, Dr. Azzeddin R. Elhawej and Dr. Khalil Suleiman 2004–2006)

In 2004, areas in north-western coastal areas that had not been covered by the study by the Russian Selkhozprom Export Company were studied. These areas extend from the western part of Al Ujeilat to Ras Ejder, and also receive precipitation ranges less than 200 mm per year. The US Soil Taxonomy system applied, and pedological soil maps and interpretive soil maps were derived at a scale of 1:50,000.

In 2006, areas such as Zuwara, Sania Oum Mansour and SawaniZrir, which are also located in the north-western coastal areas were studied. In addition, the US Soil Taxonomy system was used to define soil characteristics and furthermore, the pedological soil maps and interpretive soil maps have been obtained at a scale 1:50,000 (National Consultative Office and Public Water Authority 2006).

2.3.3.7 Soil Resources studies for the planting of (100) thousand hectares of irrigated wheat in southern part of Libya (2005)

In 2005, 11 semidetailed soil survey location studies were conducted in southern parts of Libya, including Kufra, Debutats, Irun, Macknoush, Al Ariel, Al Burjooj, Wadi Al Nashwa, SararAl Qatousa, Al Majdoul, Otlalab and west Awanat. These studies were carried out by the Astrolabe office (Libyan Consulting Engineers) in cooperation with a number of soil mapping experts at the University of Tripoli, headed by Dr. Khaled ben Mahmoud, Dr. Azzeddin R. Elhawej and Dr. Khalil Suleiman). In addition to this, the US Soil Taxonomy system was used in preparing soil classification maps and interpretive soil maps at a scale of 1:50,000 (Astrolabe Office "Consulting Engineers" and the Department of Agricultural and Pastoral Development 2005).

2.3.3.8 A Study of Soil Resources by the UNDP Office in Tripoli (2002)

The detailed soil survey study for the farms was conducted in the old city of Ghadames. The US Soil Taxonomy system has been used to describe soil features and properties. The pedological soil maps and interpretive soil maps were derived at a scale 1: 50,000 (UNDP and Agricultural Research Center, in cooperation with a number of soil mapping experts at the University of Tripoli, headed by Dr. Khaled ben Mahmoud, Dr. Azzeddin R. Elhawej and Dr. Khalil Suleiman and 2002).

2.3.3.9 Other Soil Resources Studies

There are several other studies of soil for different areas and multiple agricultural projects conducted by various bodies such as the soil section of the General Authority for Water, as well as the Hydro project and Energo Project companies from Jugoslavija. There are also studies that include sporadic areas such as Tawergha, Krarim, Tomina, Ad-Dafiniyah, Wadi ka'am, Zliten, Bin Ghashir, Shakshouk, Al Josh, Tukuk, south Jumayl, Al Urban (Hydroprojekt 1974) and Derj and Ghadames (Engineering Consulting Office for Utilities 1996 in cooperation with two soil mapping experts at the University of Tripoli, headed by Dr. Khalid ben Mahmoud and Dr. Azzeddin R. Elhawej. The purpose of most of these studies was to determine the suitability of the soil for agriculture and the establishment of agricultural projects, the capability and productivity of land, the depth of the hardpans layer and its presence, the problem of dissolving salts, and how to reduce the problem by reclaiming the salt.

A Study of Soil Resources by ACSAD (2005)
In 2005, the Arab Centre for the Studies of Arid Zones and Dry Lands (ACSAD) cooperated with the department of soil at General Water Authority (GWA) of Libya and with Dr. Khaled ben Mahmoud. One of the main goals of this cooperation was the derivation of soil and terrain Database (SOTER) and soil classification maps of Libya at scale 1: 1 M (ACSAD and GWA 2005) (will be discussed in Sect. 2.5).

Mapping of Natural Resources for Agricultural Use and Planning Project (LIB00/004) Studies (2001–2010)
In 2001, the Project of Mapping Natural Resources in Libya was launched, and through this project, a number of goals have been accomplished (LIB00/004 2007). These goals include:

- Digitization of all available soil maps resulted from previous studies.
- Establishment of Land Information Database (Land Resource Information Management System in Libya (LRIMS)).
- Derivation of Digital Land / Vegetation Cover Map of Libya. The achievements of the Project of Mapping Natural Resources will be discussed in Sect. 2.5.

2.4 Mapping the Soils of Libya (Available Soil Pedological and other Interpretive Soil Maps)

Extensive soil studies have been carried out in Libya for the last five decades. Emphasis was given mainly to the northern part of Libya, to the coastal zone, and to only some small, scattered areas in the southern desert. The present soil survey reports and maps differ in their contents, types of maps, scale of mapping, classification systems used, methods of soil analysis, criteria which the interpretation of data is based on, etc. The main problem that lies behind these variations is not how to convert the taxonomic units from one soil classification system to another, but how the results of the soil properties can be compared. Therefore, a careful analysis and evaluation of the data in these soil survey studies is needed. Another major problem facing the use of available data is that some of these soil survey reports do not include soil classification maps; they contain only interpretive soil and land maps (Ben Mahmoud et al. 1984; Ben Mahmoud 1995, 2001).

2.5 Research Projects and Achievements

As indicated above, a large number of soil studies have been conducted in Libya since 1970 to identify and classify soil resources of promising areas in the country for optimum land-use and to specify scientific and economical ways for utilizing soil and water resources in agriculture development and research activities. Soil research in soil classification and mapping has expanded since 1980. The focus in the early stages was on land classification for agricultural use and cartography. Later, in the 1990s, multiple land evaluation occurred to assess the land potential for general land uses such as suitability for crops and suitability for irrigation, dryland farming system, rangeland, etc. The main achievements and outputs until 2019 are as follows.

2.5.1 Achievements in the Field of Soil Suitability Mapping

2.5.1.1 Mapping of Soil Suitability for Agricultural Crops Using Productivity Index Rating

This Productivity Rating Index is a parametric system that was adapted and developed for Libyan soils by Prof. Khaled Ben Mahmoud in 1995. Khaled Ben Mahmoud applied the parametric land evaluation system to land suitability classification for many agricultural crops in Libya (Ben Mahmou 1995). This Rating Index has identified 11 land attributes to calculate soil suitability for agricultural crops (Eq. 2.1).

Productivity Rating Index (PRI) = A1*A2*A3*A4*A5*A6*A7*A8*A9*A10*A11.

where A1 = soil texture, A2 = soil calcium carbonate, A3 = soil depth, A4 = soil reaction, A5 = soil organic matter, A6 = soil salinity, A7 = soil slope, A8 = soil erosion, A9 = internal soil drainage, A10 = water table, A11 = exchangeable sodium percentage. Land variables in this system are given scores from 0 to 1, depending on the effect of the parameter on agricultural production in Libya.

2.5.1.2 Mapping of Soil Suitability for Agricultural Crops Using GIS Techniques

The FAO framework for land suitability for agriculture crops was first developed in the north-east of Libya (Nwer 2005). The model of land suitability was constructed using Geographic Information System (GIS) capabilities and modeling functions to accommodate barley, wheat, maize "corn" and sorghum in the north-east of Libya. Soil, climate, erosion hazards, and slope were integrated in the GIS environment as information layers and then overlaid to produce overall land suitability for each crop (Nwer 2005).

In 2010, Elaalem adapted fuzzy systems, including Fuzzy AHP and Ideal Point approaches to the FAO framework for land suitability for agriculture crops such as barley, wheat, maize in Jafara Plain of Libya. Many relevant soil and landscape criteria were identified and encoded in the GIS environment and their weights specified as a result of discussions with local experts. Fuzzy AHP and Ideal Point approaches used to drive land suitability maps for the selected crops. Elaalem concluded that the two fuzzy approaches have shown their ability to explore the uncertainties associated with describing the land properties in the Jafara Plain of Libya (Elaalem 2010).

2.5.2 Mapping of Natural Resources for Agricultural Use and Planning Project Achievements

2.5.2.1 Derivation Soil and Terrain Database (SOTER) Map of Libya

SOTER databases provide data on key soil and terrain properties that offer relevant input to agro-environmental applications such as food projection studies, climate studies, land evaluation, or hydrological catchment modeling. In addition to this, SOTER aims to establish a global Soils and Terrain Database, at scale of 1:5,000,000, containing digitized map units and their attribute data in standardized format.

In this study, all the spatial data related to soil, and lithology and topography were encoded, analyzed, and processed. Then, various SOTER units in Libya were determined and mapped, using SOTER system schemes. 136 Soter units were recognized (105 Soter units have a pedagogical basis or soil formation, and 31 Soter units have not pedagogical basis or non-soil formation). As the SOTER units were defined, all the nonspatial information (Terrain, Terrain Components, Soil Components) as well as plant cover, land use, meteorological data, analysis procedures, and map sources, were fitted in special forms.

The Relation Database Management System (RDBMS) was used to enter, edit, process and SOTRE data in the computer, using the Geographical Information System (GIS). The spatial information (maps) were combined with the nonspatial information to produce digital SOTER units at scale 1: 1 M (Fig. 2.13).

The SOTER map of Libya at scale 1:1 M was compiled by the cooperation of the Arab Centre for the Studies of Arid Zones and Dry Lands (ACSAD) with the Department of Soil at General Water Authority (GWA) of Libya. This map is in addition to the SOTER Maps of the north-east and north-west of the country prepared at scale of 1:250,000 (ACSAD and GWA 2005; Ben Mahmoud 1997, 1998, 2001).

2.5.2.2 Derivation Digital Soil Map of Libya

Preparation of a digital soil map of the country at a scale of 1:1 M was compiled in 2005 by the cooperation of ACSAD with the department of soil in GWA. The US Soil Taxonomy system was applied to produce a digital soil map of Libya (ACSAD and GWA 2005). Figure 2.14, shows Soil Map of Libya at scale 1: 1 M.

2.5.2.3 Derivation Digital Land/Vegetation Cover Map of Libya

The primary data used in this study were Landsat data with 30 m spatial resolution. The visual interpretation was applied to analyze satellite imagery data. The results showed that the cultivable areas were estimated at only 2.0 million hectares, which is slightly over 1% of the total area (1.3%). Also, the majority of the cultivated and/or rangelands are located along the northern zone. At present, the areas that are under irrigation are at about 610 thousand hectares. These areas include large scheme projects, settlements, and smallholder farms (LIB00/004 2007). Figure 2.15 shows Land Vegetation Cover classes in Libya.

2.5.2.4 Establishment Land Resource Information Management System (LRIMS) in Libya

Before 2001, none of the soil maps was in digital formats. In 2001, the Libyan Land Resources Database and Information Management System was initiated. This occurred through a project entitled Mapping of Natural Resources for Agricultural Use and Planning (LIB00/004 2007), which was implemented by FAO in cooperation with a number of Libyan experts. This project recognizes all the natural resource data in a geographically referenced computerized database, including Soils.

2.5.3 Academic Achievements in Soil Genesis, Classification, and Mapping

As mentioned in Sect. 2.3, the MSc program in the field of soil genesis, classification, and mapping was developed in some agricultural facilities in Libya such as faculty of agricultural in Tripoli, Libya, and therefore different soil scientific topics were covered. Table 2.2. shows some of the master's thesis topics applied by Libyan students in the area of soil genesis, classification, and mapping within the faculty of agricultural in Tripoli, Libya.

In addition to this, because the Ph.D. program is still not active in Libya, as mentioned above, the Libyan government granted a number of Libyan graduate students' scholarships in the field of soil genesis, classification, and mapping.

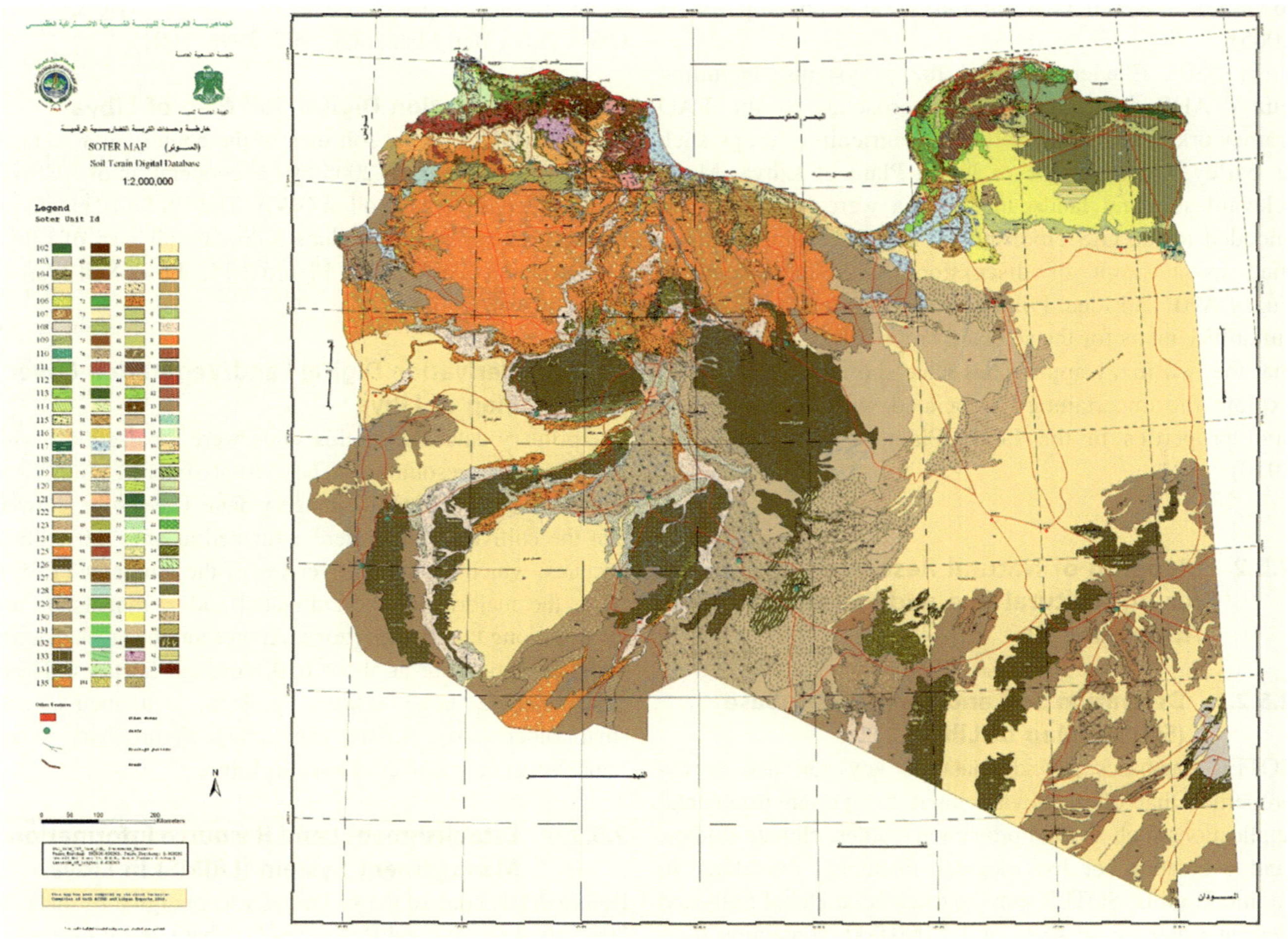

Fig. 2.13 Soter unit map of Libya (*Source* ACSAD and GWA 2005)

These scholarships applied to students studying in international universities. Table 2.3 shows some of the Ph.D. thesis topics applied by Libyan students in the area of soil genesis, classification, and mapping in some international universities.

In addition to these student projects, a few books were published by Libyan soil science experts in the area of soil genesis, classification, and mapping, and most of these books were written in the Arabic language. For example, a comprehensive literature review regarding Libyan soils can be found in the flowing books: Soil study in the field (1984) by Khaled ben Mahmoud and Adnan Ajendeel, Libyan Soils (Their Genesis Classification, Properties and Agricultural Potential) and Towards a National Strategy for the Sustainability of Natural Resources and Enhancing Food Security in Libya which were written in 1995 and 2013, respectively, by Prof. Khaled Ben Mahmoud. Furthermore, more than 500 scientific papers were accomplished by Libyan soil science researchers in different branches of soil science, and they are written in Arabic or English languages and published in national or international journals. Table 2.4. shows the examples of what has been accomplished in the area of soil genesis, classification, and mapping papers which were published in national or international journals or conferences.

2.6 Conclusions

The soil survey reports and maps created over the past decade differ in their contents, types of maps, scale of mapping, classification systems used, methods of soil analysis, criteria on which the interpretation of data based, etc. The major soil classification systems used in these reports

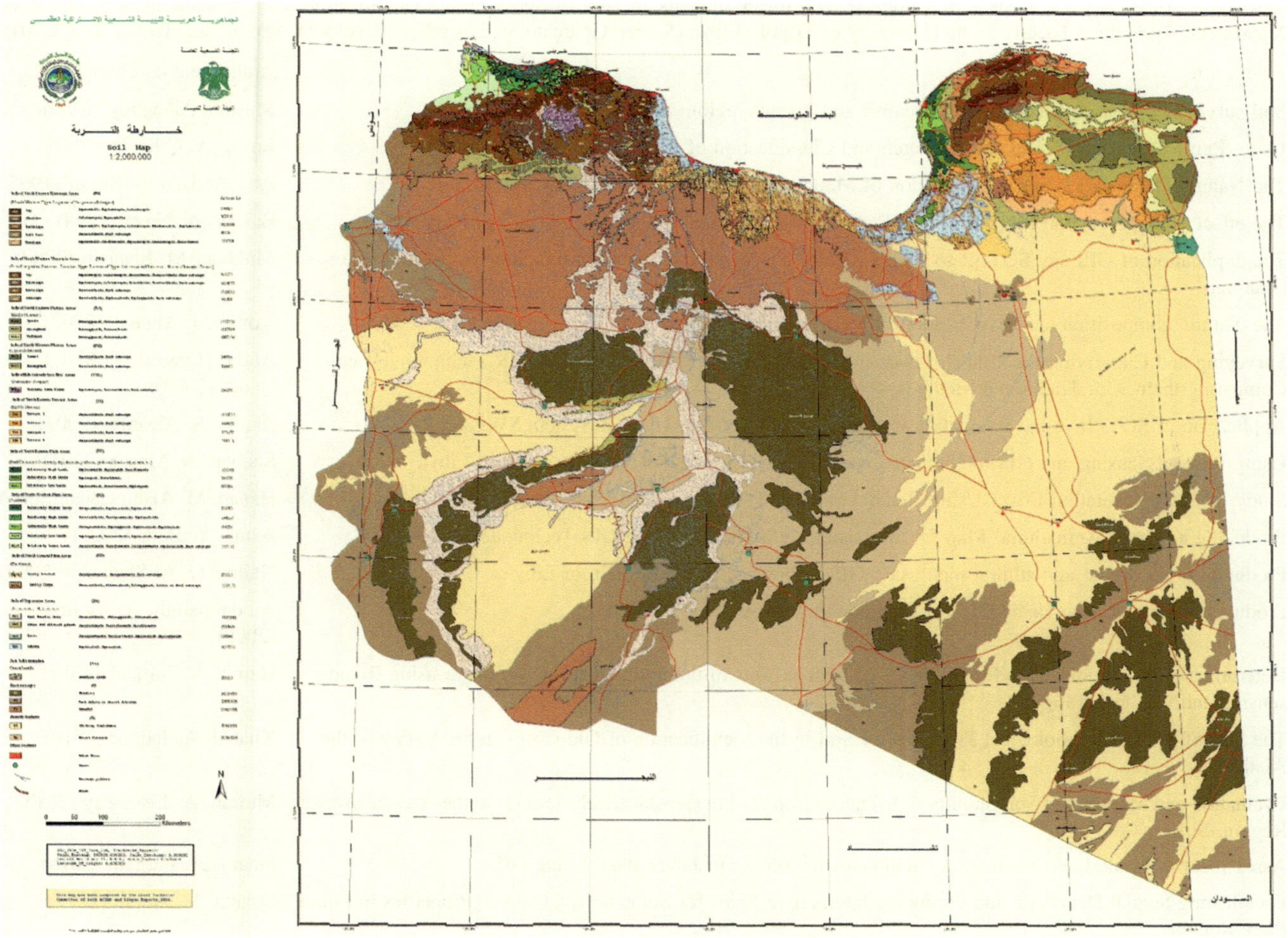

Fig. 2.14 Digital soil map of Libya (*Source* ACSAD and GWA 2005)

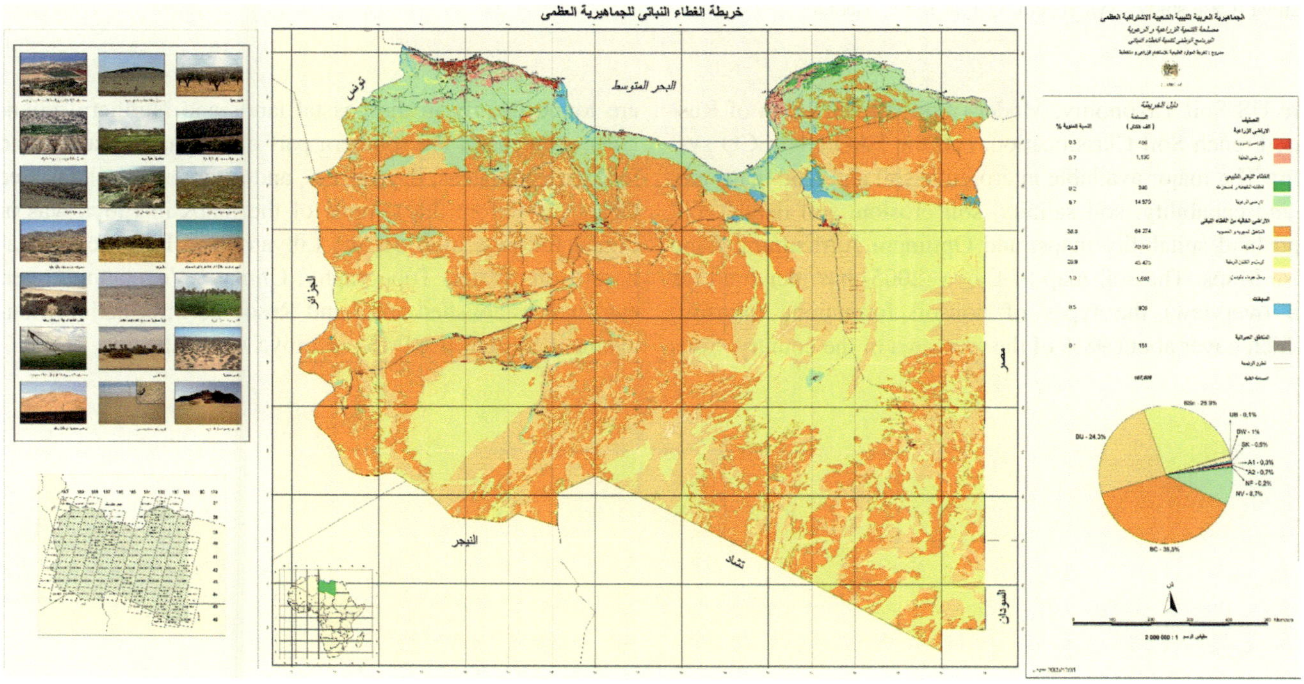

Fig. 2.15 Land vegetation cover classes in Libya (*Source* LIB00/004 2007)

Table 2.2 Some master's thesis topics applied by Libyan students in area of soil genesis, classification and mapping in Soil and Water Department, Agriculture Faculty at the University of Tripoli, Libya (*Source* Graduate Studies office, Faculty of Agriculture, Tripoli, Libya 2019)

Thiess title	Author and date
Soil survey and classification of Al Ghardabya and Swawa region	Mustafa Alhagagy (1989)
Using Remote Sensing Images in the Search and Classification of Agricultural Soils	Jamal. A. Qiliwan (1994)
The Nature, Composition and Classification of Mountain Soil in Gharyan, Yafran and Tarhuna	Abu Abdalla. S. Sherif (1995)
The effect of slope on Soil Morphological, Physical and Chemical properties in Gandoba Region	Basher. A. Nwer (1998)
The application of GIS and SOTER to Produce Soter Units Maps for an Area Extended from Tripoli to Gharyan	Mukhtar. M. Elaalem (2000)
The Nature, composition and classification of Vertisols in Libya	Yousef. F Algalaz (2002)
Surveying and Classifying the Otlalab Lands in Kufra Using Remote Sensing and GIS Technologies and Comparing them with Traditional Methods	Abdul Hameed. S. Abdul Qader (2003)
Application of SOTER in the Production of Interpretive Soil Maps Using a System SWEAP	Jamila. S. Alearabi (2003)
Using Remote Sensing and GIS to Monitor Land Degradation in Bir Koca, Jeffara Plain, Libya	Shaban. A. Mnsur (2003)
Study of Land Vegetation Cover Using Remote Sensing Technique in Garabulli Area	Hanan. M. Alshuwshan (2006)
Producing a Digital Agricultural Map for the popularity of Tripoli Using GIS Technique	Asma. Y. Badr (2009)
Production of a digital agricultural map using GIS techniques in Jafara Plain of Libya	Naima. O. Al-Shamekh (2009)
Producing a Digital Map Using GIS for the popularity of Al Jabal Al Akhdar	Aibdalmatlub. A. Alqirbae (2009)
Determination of the spread and interference of Sebkha soils in the north-west region, Libya using Remote sensing and GIS technology	Hamdi A. Zurqani (2010)
The Use of Micro-Morphological Properties of Soil in the Identification of Old Crops in the Valley of the Alathl	Khaled. A. Jaddor (2010)
Evaluation and Mapping of Soil Salinity of Irrigated Crops Using Geostatistical Methods in the Area of Swawa Region, Sirte	Muftah. A. Elsewedy (2015)
Assessment of Some Land Evaluation Techniques for Barley in Jafara Plain Using GIS	Fatama.A. Furgani (2016)
Establishing Spatial Data Base and Producing Interpretive Maps for Some Soils Chemical Properties in Jafara Plain Region	Osama. M. Anbia (2017)
Using of Remote Sensing and GIS in Detecting Land Cover Change inKhoms Area Libya (1987–2015)	Mohamed M. Ben Amara (2017)
Application of Geostatistical Analysis in Mapping some of Soil Chemical Properties for Eastern Region of Libya (Case study: Ayn Hizam, Qaryat Batth, Taknis)	Asma. A. Elgmati (2020)

are US Soil Taxonomy, Modern Soil Classification of Russia, French Soil Classification, and the FAO/UNESCO system. The major available interpretive soil and land maps are: l and capability, soil salinity, soil erosion, soil depth, soil, and land suitability maps, and Optimum Agricultural Land Use Maps. The soil map of Libya (2005) has identified (as an overview) the types of nonsoil formations (nonsoil) which cover about 98% of the total area of the country. Soils are found mainly in the coastal lands and Northern mountains, whereas in the southern part of Libya, they are found in depression areas, dry valleys, and sedimentary plains (see Chap. 3). Before 2001, none of the maps in Libya was in digital formats. In 2001, the Libyan Land Resources Database (Soil, Water, Topography, Climate, Vegetation cover, etc.) was established and Land Resource Information Management System (LRIMS) in Libya was initiated.

Table 2.3 Some of the Ph D. Thesis topics applied by Libyan students in the area of soil genesis, classification, and mapping in some international universities

Title	Author and date	University and Country of study
Modeling Soil Erosion in north-west Libya	Abu abdalla. S. Sherif, (2004)	Reading University, UK
The application of land evaluation technique in the north-east of Libya	Basher. A. Nwer (2005)	Cranfield University, Silsoe, UK
The Application of Land Evaluation Techniques in Jafara Plain in Libya using Fuzzy Methods	Mukhtar. M. Elaalem (2010)	Leicester University, UK
Assessing Land Degradation and Land Use in the Libyan Al Jabal Al Akhdar Region	Yousef. F Algalaz (2013)	Sheffield Hallam University, UK
An evaluation of Land Degradation and Desertification in the Jeffara Plain, Libya	Shaban. A. Mnsur (2014)	Sheffield Hallam University, UK
Fuzzy- GIS Development of Land Evaluation for Agricultural Production in north-west Libya	Nagib M. Froja (2013)	Heriot-Watt University, UK
Integrated Model of an Early Warning System (EWS) of Desertification in Libya	Azalarib S. Ali (2019)	Cranfield University, Silsoe, UK
Application of Geospatial Technologies for Land Use Analysis and Soil Science Education	Hamdi A. Zurqani (2019)	Clemson University, USA

Table 2.4 Some of soil genesis, classification and mapping papers published in journals or conferences

Title	Author and date	Name of paper Journal or conferences
Nature and properties of soils of Fezzan area of Libya	Ben Mahmoud K. R. and Abdelgawad (1993)	Journal of the Academy of Scientific Research
Soils in Libya and salinity problems	Ben Mahmoud K. R (1998)	The Libyan Journal of Agriculture, Tripoli—Libya
Land Degradation and Desertification in Libya	Ben Mahmoud K. R et al. (2000)	Dubai International conference on desertification
A Comparison of Fuzzy AHP and Ideal Point Methods for Evaluating Land Suitability	Mukhtar. M. Elaalem (2011)	Transactions in GIS, 2011, 15(3): 329–346
Assessment of spatial and temporal variations of soil salinity using remote sensing and geographic information system in Libya	Hamdi A. Zurqani et al. (2012)	International Proceedings of the 1st Annual International Conference on Geological and Earth Sciences, Singapore, 3–4 December 2012
Land Suitability Evaluation for Sorghum Based on Boolean and Fuzzy—Multi-criteria Decision Analysis Methods	Mukhtar. M. Elaalem (2012)	International Journal of Environmental Science and Development., vol. 3, No. 4,
Performance of Supervised Classification for Mapping Land Cover and Land Use in JafaraPlain of Libya	Mukhtar. M. Elaalem et al. (2013)	International Proceedings of Chemical, Biological and Environmental Engineering IPCBEE, vol. 55 (2013), IACSIT Press, Singapore
A Comparison of Parametric and Fuzzy Multi-Criteria Methods for Evaluating Land Suitability for Olive in Jafara Plain of Libya	Mukhtar. M. Elaalem (2013)	Elsevier and included in ScienceDirect, vol. 5, pp. 405–409
The Use of Remote Sensing and Geographic Information System for Soil Salinity Monitoring in Libya	Basher. A. Nwer et al. (2013)	GSTF Journal of Geological Sciences Vol. 1 No. 1
Soil Productivity Rating Index Model Using Geographic Information System in Libya	Basher. A. Nwer et al. (2013)	GeoTunis conference, Tunis Tunisia
Techniques to Control Wind Erosion in Libya	Basher. A. Nwer (2014)	Revue des Arides - NuméroSpécial- n° 36

(continued)

Table 2.4 (continued)

Title	Author and date	Name of paper Journal or conferences
Spatial Variability of Some Soil Chemical Proprieties in Jeffara Plain, Libya (Case Study: Tripoli, Wadi Almjainin and Bin Ghashir)	Mukhtar. M. Elaalem (2017)	The Libyan Journal of Agriculture, volume (22): No. (1): 2017: 19–34
Predicting the classes and distribution of salt-affected soils in north-west Libya	Hamdi A. Zurqani et al. (2018)	Communications in Soil Science and Plant Analysis. 2018, 49, 689–700
A Review of Libyan Soil Databases for Use within an Ecosystem Services Framework	Hamdi A. Zurqani et al. (2019)	Land 8, no. 5 (2019): 82

References

ACSAD and GWA (2005) SOTOR, and Soil Map of Libya at scale 1:1M, and SOTOR Map of NW and NE of Libya at scale 1:250000

Astrolabe Office "Consulting Engineers" and the Department of Agricultural and Pastoral Development (2005) Technical and economic feasibility study for the project of agriculture (100) thousand hectares irrigated wheat. Semi-detailed soil studies for 11 sites in southern Libya

Ben Mahmoud KR (1995) Libyan Soils (Their Genesis, Classification, Properties and Agricultural potentials) NASR, Tripoli, Libya, p 615 (in Arabic)

Ben Mahmoud KR (1997) Application of SOTER in Libya. REN. FAO. Cairo. Egypt, A Technical Report, Regional Soil Office

Ben Mahmoud KR (1998) Soils in Libya and salinity problems. J Agric Res. ARC. Tripoli - Libya, pp 43–151. (in Arabic.)

Ben Mahmoud KR (2001) SOIL RESOURCES IN LIBYA (National Report). Workshop on SOTER Database. Sub-Regional Office for North Africa. Rabat, Morocco, 12–16 November 2001

Ben Mahmoud KR, Salem O, Suleiman K (1998) Soils and water resources in Libya. Submitted to FAO/SRONA, Tunisia

Ben Mahmoud KRJ, Ajendeel AR (1984) Soil Study in the Field. Tripoli University publication (in Arabic), p 272

Export S, Expedition SE, USSR (1980) Soil studies in western zone, eastern zone and central zone of Libya. Ministry of Agriculture and Land Development. Tripoli, Three separate volumes

Engineering Consulting Office for Utilities (1996) Ghadames Project. Volume II: Derj and Ghadames Soil and water survey

El-azzabi (1999) The National Agriculture research system of Libya. In the NARS of West Asia and North Africa. Editor: Joseph Cosas (ICARDA). p 69–6

Elaalem M (2010) The Application of Land Evaluation Techniques in Jeffara Plain in Libya using Fuzzy Methods. Published PhD thesis, Leicester University

Farler JL (2006) Italian Agricultural Settlement in Libya (Tripoli Area). Publications of the Libyan Jihad Center for Historical Studies, Translated by Abdul Qader Al Muhaishi

GEFLI (1971–1978) Soil Survey and mapping in different locations in Libya (exploratory, semi-detailed and detailed studies, about 20 locations)

Habert P (FAO) (1964) The soils of northern Cyrenaica. Soil Survey Report (Scale of 1:100000)

Hydroprojekt (1972–1974) Soil Survey and mapping in different locations in Libya (exploratory, semi-detailed and detailed studies, about 6 locations)

LIB00/004 (2007) Mapping of Natural Resources for Agricultural Use and Mapping. (GAA, FAO & UNDP). Unpublished data

National Consultative Office and Public Water Authority (2005–2006) Completion of a technical study on Soil survey and classification of the north-western coastal areas of Libya (Zuwara to the Tunisian border, Zuwara, Sania, Oum Mansour, and SawaniZrir). Report and maps at scale 1: 50000

Nwer BA (2005) The application of land evaluation technique in the north-east of Libya. Published PhD thesis, Cranfield University, Silsoe

REGWA (1993) Soil Survey and Mapping in Al magdoud, Tjehi, and Al-gatroun area in south of Libya. Semi-detailed mapping

REGWA (2005) Soil Survey and Mapping in Al Sarah area in south of Libya. Semi-detailed mapping

Stewart JH (1960) Land and water Resources of Tripolitania (report and maps, scale of 1: 100000)

Tele Kart (1994) Semi detailed soils study of Al Necat Al Khames. Final Report

Dr. Mukhtar M. Elaalem is an associate professor of soil pedology and head of the Department of Soil and Water Sciences, Faculty of Agriculture, University of Tripoli, Tripoli, Libya. He got a Ph.D. degree in Geographic Information System at Leicester University, UK in 2011. His Ph.D. thesis is on "The Application of Land Evaluation Techniques in Jafara Plain in Libya using Fuzzy Methods", and his research interest is soil pedology, geographical information science, natural resources applications of GIS (especially land evaluation and soils studies), modeling spatial uncertainty, especially employing fuzzy logic methods to land evaluation and soils studies. He teaches various courses, including introduction to soil science, soil survey; genesis and classification, and Libyan soils. He attended many local, regional, and international meetings, workshops, training programs, conferences on land resources, and applications of GIS and remote sensing.

Dr. Hamdi A. Zurqani is one of the faculty members at the Department of Soil and Water Sciences, Faculty of Agriculture, University of Tripoli, Tripoli, Libya. Dr. Zurqani is a recognized expert as a result of his internationally acclaimed work in the areas of environmental information science, remote sensing, land evaluation, sustainability, pedology, and soil science education. He has conducted research across the world, including the United States of America and Africa. Dr. Zurqani is a distinguished soil scientist with a wide range of scientific and working experiences in Libya and abroad. He received his M.Sc. (2010) from the University of Tripoli, Tripoli, Libya and Ph.D. (2019) from Clemson University, Clemson, SC, USA. His major research and teaching activities at the University of Tripoli have focused mainly on soil genesis and classification and the environmental information sciences (Remote Sensing and Geographic Information System). He has published broadly in many journals (e.g., Nature "Scientific Reports", Geoderma; International Journal of Applied Earth Observation and Geoinformation; Journal of Geological Sciences; Land; Frontiers in Environmental Science; Communications in Soil Science and Plant

Analysis; and others). Dr. Zurqani is a member of the Editorial Board for Remote Sensing (MDPI) Journal, counseling outcome and research evaluation. He also was appointed to serve as a Guest Editor for the Special Issue "Applications of Remote Sensing in Earth Observation and Geo-Information Science'. In addition, Dr. Zurqani conducted peer-review for many journals including Journal of Environmental Informatics, Applied Sciences, SN Applied Sciences, Remote Sensing, Heliyon, Geosciences, Land, Water, Agronomy, Agriculture, Sustainability, Arid Land Research and Management, International Journal of Environmental Research and Public Health, Natural Hazards, and Conference of the Arabian Journal of Geosciences. He is also one of the authors of the lab manual entitled, "GIS Exercises for Natural Resource Management". Dr. Zurqani has been the recipient of numerous awards and honors: Recipient of Douglas R. Phillips Award for Graduate Students, Department of Forestry and Environmental Conservation, Clemson University, April 12, 2019; the First Place Best Judged Poster (CAFLS) at the GRADS 2019: Clemson Student Research Forum on April 4, 2019; the Second Place Poster at the 11th Clemson Biological Sciences Annual Student Symposium, April 6, 2019; the Second Place Best Judged Poster at the Clemson Student Research Forum on April 4, 2018; and the Third Place Poster at the 9th Clemson Biological Sciences Annual Student Symposium, February 25, 2017. Dr. Zurqani conducts cutting-edge research in the field of environmental information science, remote sensing, land use management/planning, change detection of landscape degradation, and geographic information system (GIS) models. He has focused on his research efforts on the development of new technologies in the field of environmental information sciences, geo-intelligence (advanced geo-information science and earth observation, machine and deep learning, and big data analytics), remote sensing, land evaluation, pedology, land use management/ planning, monitoring and evaluating sustainable land management, change detection of landscape degradation, and geographic information system models.

Dr. Khaled R. Ben Mahmoud is a senior professor of soil pedology. He is a distinguished soil scientist with wide range of scientific and working experiences in Libya and abroad. He received his Ph.D. in 1977 from Colorado State University (USA). His major teaching activities and research are at the Department of Soil and Water, Faculty of Agriculture, University of Tripoli, Libya. He has been teaching undergraduate and graduate students in Libya and has supervised many M.Sc. and Ph.D. thesis and dissertation in Libya and abroad. He focused mainly on soil resources, land evaluation, and mapping. He contributes to soil science in Libya and abroad as the Head of the Department of Soil and Water Sciences, Faculty of Agriculture, University of Tripoli, Tripoli, Libya (1978–1983), Dean of the Faculty of Agriculture, University of Tripoli, Tripoli, Libya (1983–1986), Chairman of the Committee for Libyan Agricultural Mapping, Tripoli, Libya (1988–1990). A Research Coordinator, University of Tripoli, Tripoli, Libya (1997–2001), Member of the Scientific Council of Agricultural Research Center (ARC), Tripoli, Libya (2000–2002), Member of the Scientific Council of the College of Higher Studies at the University of Tripoli, Tripoli, Libya (2000–2002), Member of the Scientific Committee of Libya Centre for Remote Sensing and Space Sciences, Tripoli, Libya (1998–2006), A National Coordinator of the Project implemented by FAO: Mapping of Natural Resources for Agricultural Use and Planning Tripoli, Libya (2000–2006), A General Coordinator of the National Programme for the Development of Vegetation Cover, Tripoli, Libya (2004–2006). Director of the Division of Land and Water Uses, The Arab Centre of the Studies of Arid Zones Lands (ACSAD), The League of Arab States, Damascus, Syria (2007–2011), Head of the Programme: Monitoring and Combating of Desertification in the Arab World, The Arab Centre of the Studies of Arid Zones Lands (ACSAD), The League of Arab States, Damascus, Syria (2007–2011), Director General of Eddayra Co. (Agricultural and Environmental Consulting and Services), Tripoli, Libya (2011–to date). Engaged as a local and international expert to control, evaluate, approve, and accept of many soils studies and mapping which have been implemented by different foreign and local companies, agencies, and offices in different parts of Libya and abroad (1977–to date). More than 100 country reports, research and scientific papers, and other publications were published in different local and international scientific journals and symposiums (in Arabic and English). He is the author of more than 10 books that have been written in Arabic. The most important ones are: Sandy soils of Libya (1981), The study of soils in the field (1984), Principles of Soil Science for Agricultural Secondary and Technical Schools (1985), Fundamentals of Soil Science and its Relation to Plant Growth (1995), Libyan Soils (Their Genesis, Classification, Properties, and Agricultural Potentiality (1995), Fundamentals of Soil and water Science (2001), Issues on Desertification and Land Degradation in the Arab Region (2010), Towards a National Strategy for the Sustainability of Natural Resources and Enhancing Food Security in Libya (2013), and The Libyan Desert (its natural potentials and prospects for investment) in press. Finally, he was appointed as a technical advisor to the President of the Libyan Presidential Council for Agriculture, Tripoli, Libya (2018–to date).

Soil Forming Factors and Processes

3

Khaled R. Ben Mahmoud and Hamdi A. Zurqani

Abstract

Soil forming factors not only affect the course of the genesis process but also the present and future performance of the soil system, in terms of how much organic matter, air, and water and how many organisms, nutrients, and minerals it can store. This chapter discusses the soil formation factors (climate, biota, parent material, relief or topography, and time) and their impact on the soil formation processes prevailing in different regions across the country. It has become clear that the factors of soil formation in Libya are mainly influenced by the hot and dry climate (low and irregular rainfall, high temperature, with a wide range of daily and seasonal temperature fluctuations, and high wind speed), which plays an important role in the sparseness of vegetation in the region. Climate dependency dictates particular types of soil formation processes that are more dominant in Libya than in other countries and reduce—if not entirely nullify—the effectiveness of many of them. Consequently, the soils produced in Libya are similar to the soils of arid regions in the world that are undeveloped or newly developed and are typically either Entisols or Aridisols. The situation is slightly different in coastal areas (i.e., Al Jabal Al Akhdar, and some small areas in the mountains of Tripoli), where the precipitation rates are higher and the climate is milder than in other parts of the country. In these areas, the vegetation cover is more widespread, and the processes of soil formation are more active. Thus, the local topography and the parent material are the main factors that influence the formation of more developed soils in these areas.

Keywords

Arid land • Libya • Soil pedology • Parent material • Soil formation • Weathering processes • Soil horizons

3.1 Introduction

The factors and processes of soil formation have a great influence on soil properties, which determine soil profile characteristics and pedogenic pathways for the soils. The processes of soil formation at any location are complex, consisting of a mix of change, addition, loss, and transfer of mineral and organic materials. These processes are known to occur in the presence of soil forming factors, both active and passive. Soil formation processes can either work in unison, or a group of them may prevail at the same place and time to form A Soil. As soils form, distinct layers known as horizons develop over time. These horizons are distinguished from each other in physical, chemical, and biological properties, and each has distinct morphological features and a genetic relationship that connects them (Ben Mahmoud 1995). The arrangement of these horizons in a soil is known as a soil profile. Most soils have three major horizons. These are A Horizon, B Horizon, and C Horizon. Aside from these three horizons, there are also the O, E, and R horizons (Fig. 3.1).

The intensity and effectiveness of these soil formation factors differ from one geographical region to another, where each region has its own conditions that control the factors and processes of soil formation to produce distinct soils that express these variations. This clarifies the differences in the soil types in the world (soils in wet areas compared to soils in dry areas) or within the same geographical region (i.e., Libya), despite the fact that all regions are subject to the same processes of soil formation (Ben Mahmoud 1995).

K. R. Ben Mahmoud · H. A. Zurqani (✉)
Department of Soil and Water Sciences, Faculty of Agriculture, University of Tripoli, Tripoli, 13538, Libya
e-mail: h.zurqani@uot.edu.ly

K. R. Ben Mahmoud
e-mail: benmahmoud.kh@gmail.com

H. A. Zurqani (ed.), *The Soils of Libya*, World Soils Book Series,
https://doi.org/10.1007/978-3-030-66368-1_3

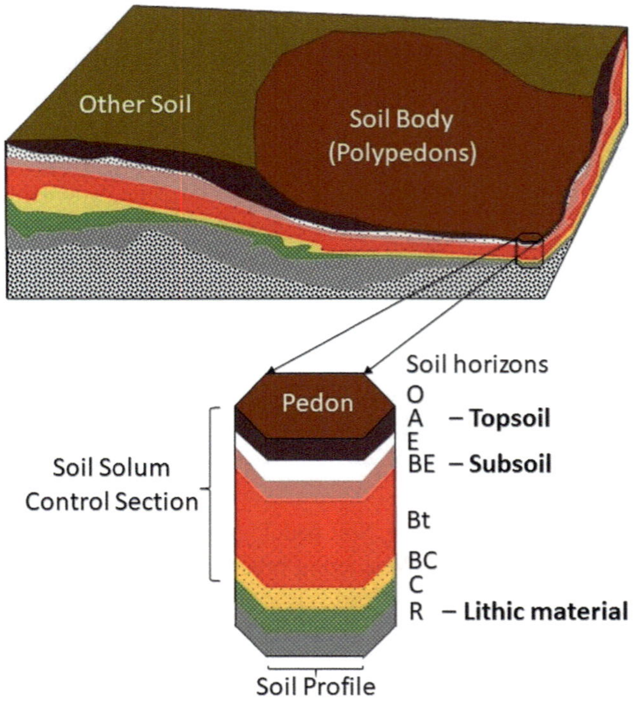

Fig. 3.1 A typical Pedon, soil solum, and the horizons of the soil profile; O Horizon—humus or organic material); A Horizon—organic accumulation in mineral soil; E Horizon—leached horizon (Elluviation); BE horizon—transitional horizon from E to B horizon; B Horizon—zone of illuviation; Bt horizon—contains illuvial layer lattice clays; BC Horizon—transitional horizon from B to C horizon; C Horizon—unconsolidated parent material; and R Horizon—lithic material; consists of slightly broken-up bedrock

3.2 Soil Forming Factors

The soil formation factors in Libya, like soil in any given region of the world, are limited to five main factors, which are climate, biota (the most important in Libya is vegetation), parent material, topography, and time. These factors have different influences that interact with each other to form the soils of Libya, causing different soil horizons to form. The following are two kinds of these factors that contribute to soil formation in Libya:

- **Active soil forming factors**: These include climate and biology (vegetation), as they are responsible for all weathering and formation processes that occur in the soil (within the soil profile).
- **Passive soil forming factors**: These include the parent material, topography, and time (i.e., the existing factors). These factors represent the mass of the earth, its various rocks, and minerals, as well as the state in which it exists in terms of shape and topography at a certain time.

It is worth mentioning here that these factors interact with one another at the same time, where they not only directly influence the soils but also have an effect on each other. For example, the climate affects the vegetation status, and vegetation cover in turn also affects the effectiveness and impact of the climate. Relief also affects the redistribution of climate (local climate). The impact of these factors on the type of soil formed in Libya varies from one place to another. In the southern regions, the factors of parent material, topography, and time are the prevailing soil forming factors, while in the Al Jabal Al Akhdar the climate and the parent material are the main responsible factors for the soil formation processes (Ben Mahmoud 1995).

3.2.1 Climate

The climate, with its various elements (i.e., precipitation, temperature, wind, and relative humidity), affects the type of soil formed wherever it is found, and it contributes directly and effectively to the weathering and the soil formation processes. Libya is subject to the desert climate that prevails in the northern part of the African continent, aside from the Al Jabal Al Akhdar and a portion of the Tripoli Mountains (i.e., Jabal Nafusa) where these areas have an arid to semi-arid climate (i.e., Mediterranean climate "Xeric") with rainfall in the winter and almost no precipitation in the summer (Zurqani et al. 2019). These areas do not exceed 2% of the total area of the country (Sharaf 1971; and Al Hajjaji 1989). Figure 3.2 shows the classification of climate zones in Libya.

The following is a brief review of the most basic elements of the Libyan climate, which have a significant influence on the types of soils that formed, and the manner in which they form, in different geographic areas of the country.

3.2.1.1 Precipitation

In fact, apart from about 3 million hectares in the northern part of Libya which receives precipitation of more than 200 mm/year and exposed to the Mediterranean climate, all the other regions have a hot and arid climate (desert climate) which is characterized by low precipitation and bad distribution (monthly and yearly) with high variability from one region to another.

It is also noteworthy to mention that the average precipitation rate varies across Libyan regions. Consequently, while the average precipitation at the coastline area and especially at the Al Jabal Al Akhdar region may reach 600 mm/year (this high precipitation can be linked to the fact that the edge of the mountain is interrupted by water from the Mediterranean Sea), in Tripoli Mountain regions, the average precipitation fluctuates from 200 to

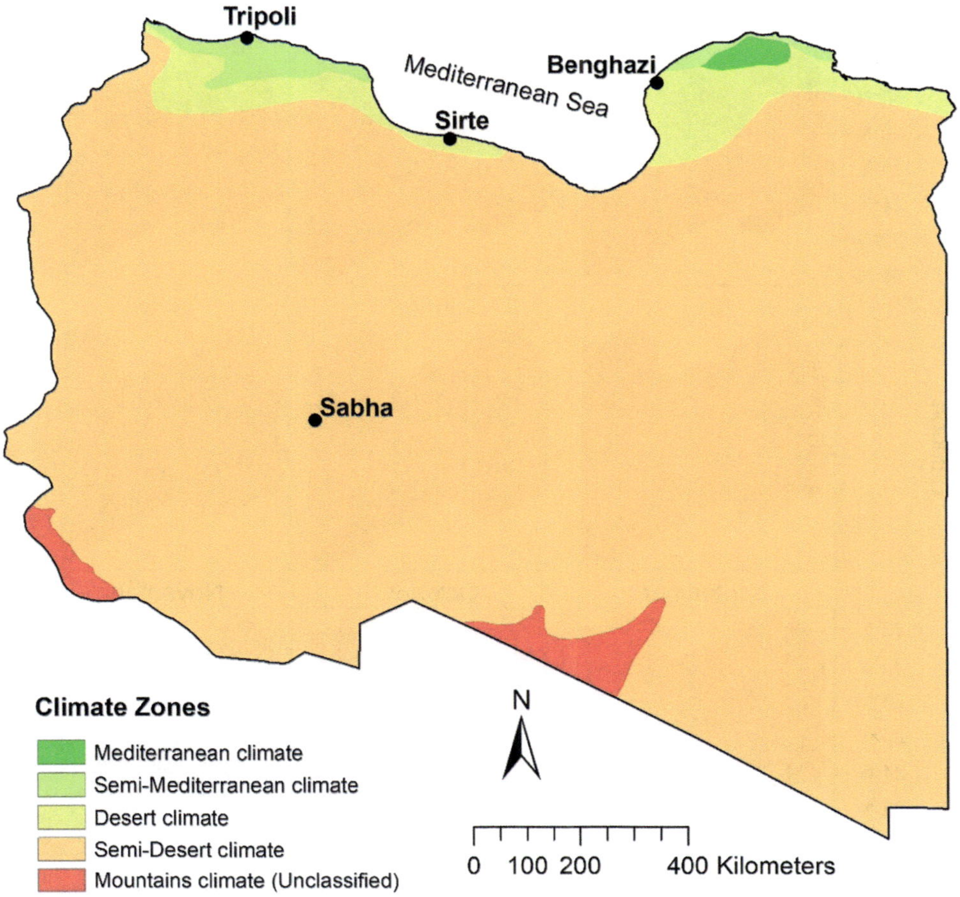

Fig. 3.2 The classification of climate zones in Libya. (Adapted from The National Atlas of 1978)

350 mm/year. Furthermore, in the southern desert regions, the average precipitation is less than 50 mm/year (Ben Mahmoud 2013; Fick and Higman 2017) (Fig. 3.3).

3.2.1.2 Temperature

Temperature is one of the most important climate elements in terms of its direct effect on the physical weathering that occurs in rocks. The temperature also has an indirect effect on the processes of soil formation, due to its influence on the type of plant life. In addition to the variability in average precipitation, the monthly average temperature for Libyan regions also shows noticeable variation. Usually, the minimum temperature is recorded in January and it increases gradually from February until July (Fig. 3.4).

According to (Ben Mahmoud 2013), the temperature difference in Libya can be summarized as follows:

- Average annual temperatures range 9–32 °C.
- The average maximum monthly temperature ranges 17–32 °C.
- Average monthly minimum temperatures range 9–23 °C.
- The highest monthly temperature ranges 30–47 °C.
- The lowest monthly temperature ranges 2–16 °C.

Despite the fact that the Libyan climate is hot and dry (i.e., desert climate) over most of the country, the geographical variation, the difference in terrain, and the direction of the coastal region are reflected in the temperature. As the climate in this region is moderated along the coastal littoral by the Mediterranean Sea,thus, the annual maximum temperature extremes in the coastal areas range 23–25 °C, while in the semi-desert areas they range 25–28 °C and more than 30 °C in the heart of the desert. On the contrary, the monthly range of temperatures is lower in the coastal areas than in the regions far from the sea. In these areas, the average monthly temperature ranges 9–11 °C in Tripoli, Misrata, and Benghazi, and it may reach 12–14 °C in the mountainous areas such as Gharyan, Yafran, Al Marj, and Shahat, while it ranges 15–18 °C in the rest of the country. As shown in (Figs. 3.2, and Fig. 3.3), for most of the months the precipitation happens to have the lowest temperatures, and that has a negative impact on the speed of the soil development process.

3.2.1.3 Relative Humidity (RH %)

The relative humidity also varies from one region to another, and this variability depends on the temperature, the altitude,

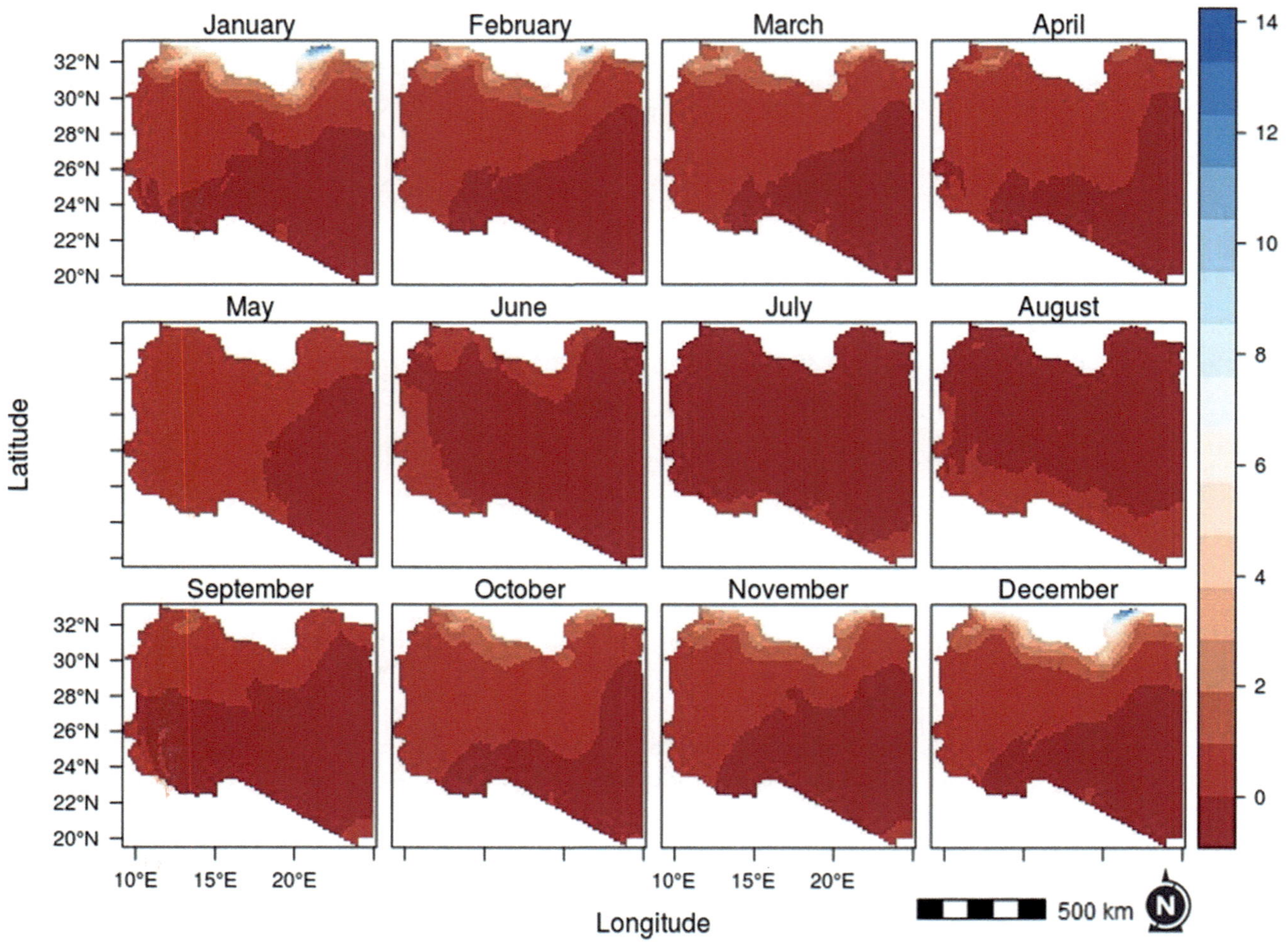

Fig. 3.3 Average Monthly Precipitation (mm) in Libya. (*Data source* Fick and Hijmans 2017)

and the distance from the sea. Thus, while the relative humidity is quite high at the coastline regions (Fig. 3.5), exceeding 80% (i.e., Al Jabal Al Akhdar), it decreases gradually to reach less than 30% in some south-desert regions of the country (i.e., Al Kufra), (The Project of Mapping Natural Resources 2006).

As a result, it can be concluded that the majority of Libyan regions have low annual precipitation (in many areas there is no precipitation), high temperature range (annually and monthly), and obvious variability in the relative humidity. Consequently, these dominant climatic conditions create soil temperature and moisture regimes distinct to these specific regions.

As many references (Ben Mahmoud 1995; Elhawej and Elaalem 2012) indicated that soil moisture regime associated with the Mediterranean climate and known as Xeric is dominant in coastline regions, where winters are cool and moist and summers are warm and dry (Zurqani et al. 2019), whereas Aridic or Torric soil moisture regime is dominant in the middle and south regions. For the soil temperature regimes, references indicated that in all northern parts of Libya, the soil temperature regime of Thermic is dominant. Meanwhile, the soil temperature regime of Hyperthermic is mostly dominant in the southern parts of the country.

3.2.1.4 Atmospheric Pressure and Wind

Atmospheric pressure and winds are among the most important climatic elements. The difference in atmospheric pressure is also known to control the movement of the winds from one region to another (i.e., the wind blows from higher pressure to lower pressure). As the wind moves across the surface of the soil, it pulls soil with it and deposits it elsewhere. Generally, Libya is influenced by high air pressure during the winter season, where almost half the northern region is exposed to a small band of low air pressure, while the rest of the country is dominated by the northeast trade winds. In the summer, as a result of the change in the low pressure that occurs on the Sahara Desert, most of the country falls within the influence of the northern trade winds (Sharaf 1971; Al Hajjaji 1989).

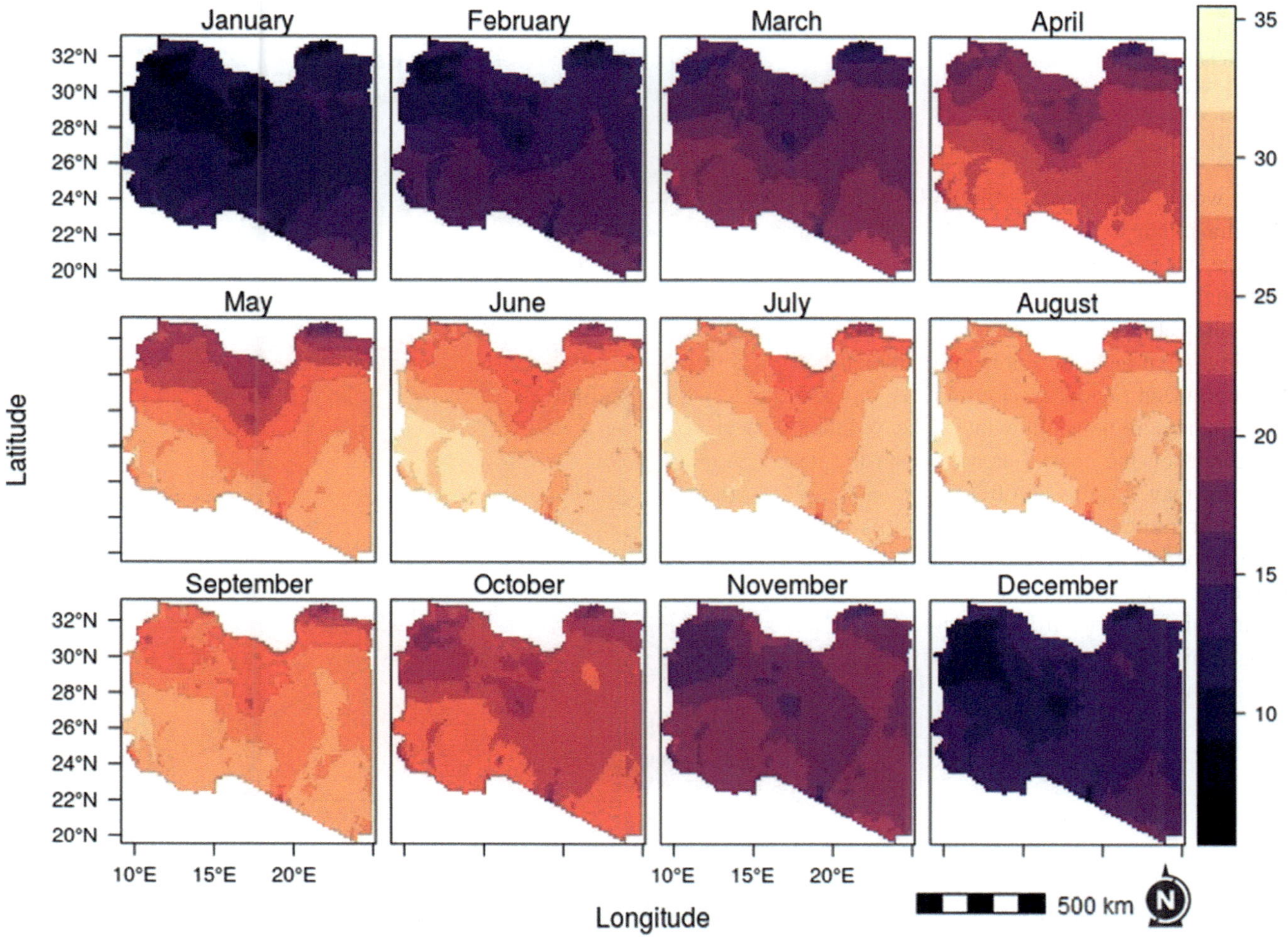

Fig. 3.4 Average monthly temperature (°Celsius) in Libya. (*Data source* Fick and Hijmans 2017)

The main distinguishing types of winds in Libya:

- **Cold wind**: This wind blows during the winter season from the north and northwest, and it is loaded with rain and increases the amount of relative humidity.
- **Southern wind**: This is hot and dry continental winds (desert winds) locally known as Qibli wind. It contributes to raising the temperature and low relative humidity, and its blows start from the first of March until the end of October in an irregular pattern. This wind causes great damage to the crops and the environment in general.

3.2.2 Biota (Vegetation Cover)

Biota or also called organisms refers to soil fauna such as earthworms and the natural vegetation cover, in which human beings have no role in its growth or cultivation. Vegetation cover is also considered one of the important soil formation factors, and it is the responsible source for organic matter in the soil. It also plays a significant role in the physical and chemical weathering processes. Moreover, vegetation cover has an indirect role in weather modification (also known as weather control).In particular, the nature of vegetation affects interception and evapotranspiration processes. Likewise, the distribution of vegetation modifies the balance between fluxes of moisture originating from the soil and those derived through canopy processes. The vegetation cover also varies in terms of its effect on soils that formed in the place, due to the difference in the amount, type, and degree of decomposition of organic matter (see Chap. 6, Sect. 6.3 for further information about the distribution of vegetation types in Libya).

In Libya, the natural vegetation cover is divided into three main parts, grass or herbaceous, shrubs, and trees (Fig. 3.6). Thus, it can be said that the amount of organic matter in the Libyan soils mainly depends on the type of the vegetation cover. For example, soils formed over the natural grass and herbs contain a greater amount of organic matter uniformly

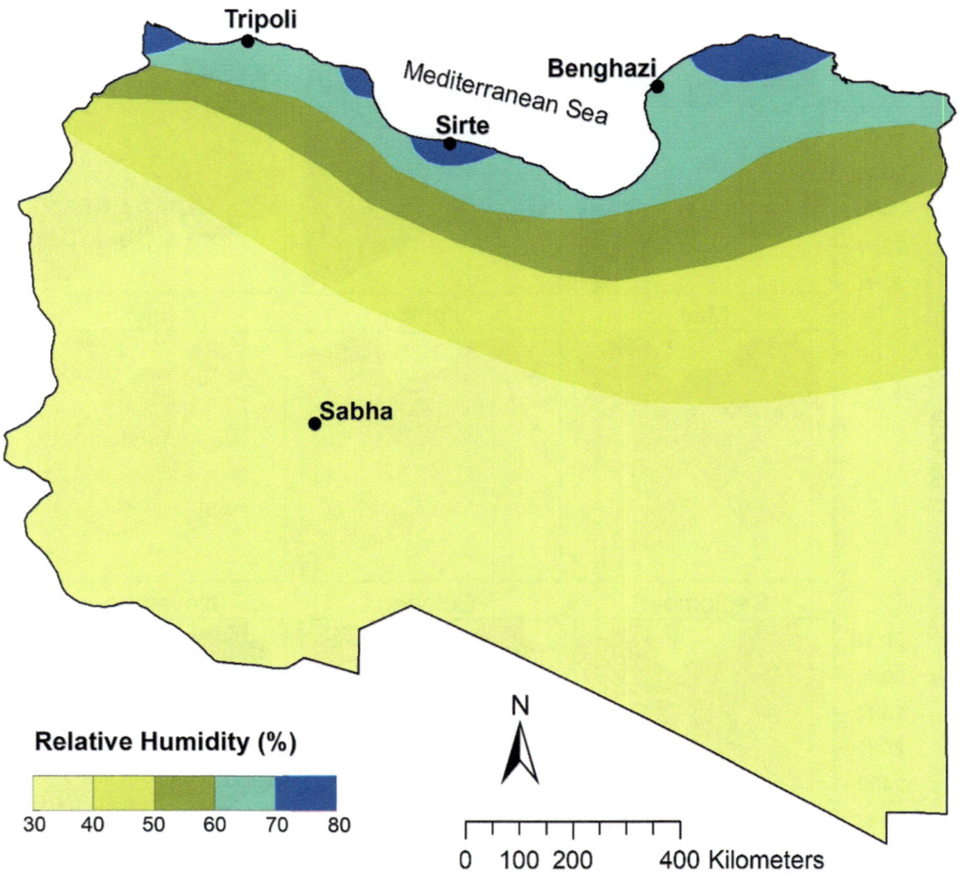

Fig. 3.5 The geographical distribution of the average annual relative humidity (percentage) in Libya. (Adapted from The National Atlas of 1978)

distributed with depth whereas, soils formed over deciduous trees contain a higher amount of organic matter than soils formed over evergreen trees.

Generally, the natural vegetation cover in Libya can be limited to Mediterranean plants, and Desert plants:

- **Mediterranean plants** include seasonal weeds, forest trees and shrubs.
- **Desert plants** include poor steppe weeds and sparse desert dwarf shrubs.

The most important types of natural vegetation in Libya and its distribution areas are shown (Table 3.1).

3.2.3 Relief (Topography)

Topography has a significant impact on soil formation as it determines runoff of water, and its orientation affects microclimate which in turn affects vegetation. In the northern region, the higher we are above sea level, the climate is moister and cooler, and the natural vegetation is denser and in much better condition. The topography and the shapes of the land surface (Terrain) also have a clear effect on the amount of water that enters into the soil, because of its impact on the surface runoff. In addition, topography has a significant impact on the depth of the water table level, especially in the lowland of the northern and southern deserts, which affects the soil formation in these regions. Figure 3.7 shows an overview map of the regional topography (topographic divisions) in Libya.

- **The plains and western high mountainous regions**: These areas extend from the Tunisian border in the west to Misrata in the east. These areas consist of two regions:
  - The coastal plains region called Jafara Plain, where the are intensive irrigated crops, in addition to rainfed crops.
  - The high mountainous region called (Nafusa Mountain), from the Tunisian land in the west to the coast of the Al-Khums city in the east, where olive, grapes, almonds, and other crops are grown, as well as some forested areas of Pistacia atlantica.
- **The eastern coastal region**: This area extends from the Libyan-Egyptian border to the city of Benghazi. It consists of a plateau that gradually increases in height until it reaches 860 m in the Al Jabal Al Akhdar region and descends toward the interior. This plateau range consists of three regions:

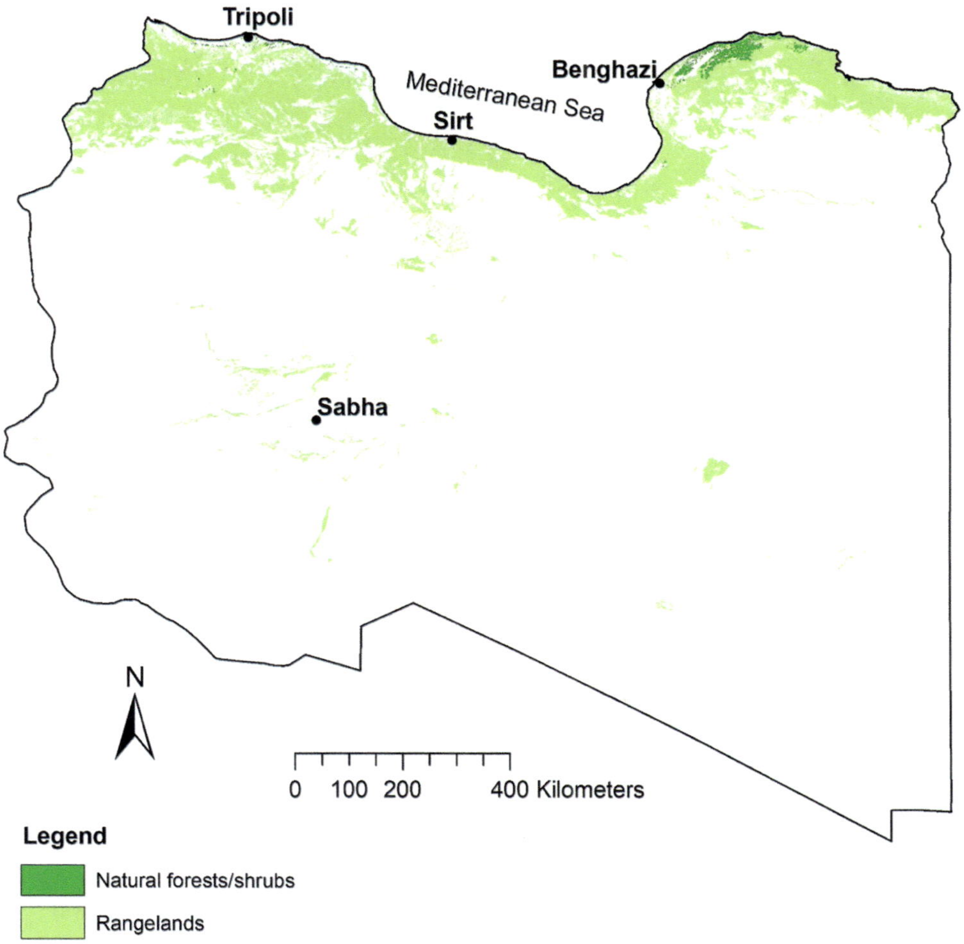

Fig. 3.6 The geographical location of the natural vegetation cover in Libya. (Adapted from LIB/00/004 2009)

Table 3.1 The most important types of natural vegetation in Libya and its distribution areas (Ben Mahmoud 1995)

Region	Vegetation type
Coastal strip areas (e.g., Castelverde (Garabulli) and Al-Khums)	Seagrass, Forest trees, Shrubs (including Rhus tripartita, Pistacia lentiscus, Calicotome villosa, and Pinus halepensis)
Northern mountain areas (e.g., Al Jabal Al Akhdar)	Dense vegetation cover from forest trees and evergreen shrubs (including Juniperus phoenicea, Pistacia lentiscus, Quercus coccifera, Arbutus pavarii, Cupressus sempervirens, etc.)
The Northern plains and the bordering desert areas (e.g., Central and Southern Jafara Plain and Sirte Plain)	Annual plant, Dwarf shrub, Retama raetam, Stipagrostis pungens, Hammada scoparia, Artemisia herbo-alba, Pituronthos Torthosus, Atriplex mollis, etc.

- **Al Jabal Al Akhdar region**: This mountain area extends from the Al Rajma region in the west to the Derna region in the east, with a length of about 400 km and a width of about 30 km. This region covers the remains of natural forests that consist of evergreen Cupressus sempervirens, Quercus coccifera, Pinus halepensis, and Juniperus phoenicea.
- **The south region of Al Jabal Al Akhdar**: This area is located along the south of Al Jabal Al Akhdar. It is a rangeland in which various shrubs and grasses grow, such as Artemisia herba-alba and Atriplex mollis, in addition to other annual and perennial plants.
- **The eastern coastal plain region**: This area includes the Benghazi and Derna plain, which are agricultural and rangeland areas.

- **The central region**: This region extends from Misrata in the west to Benghazi in the east. It is a coastal plain region with abundant saline wetlands (i.e., Sebkha),

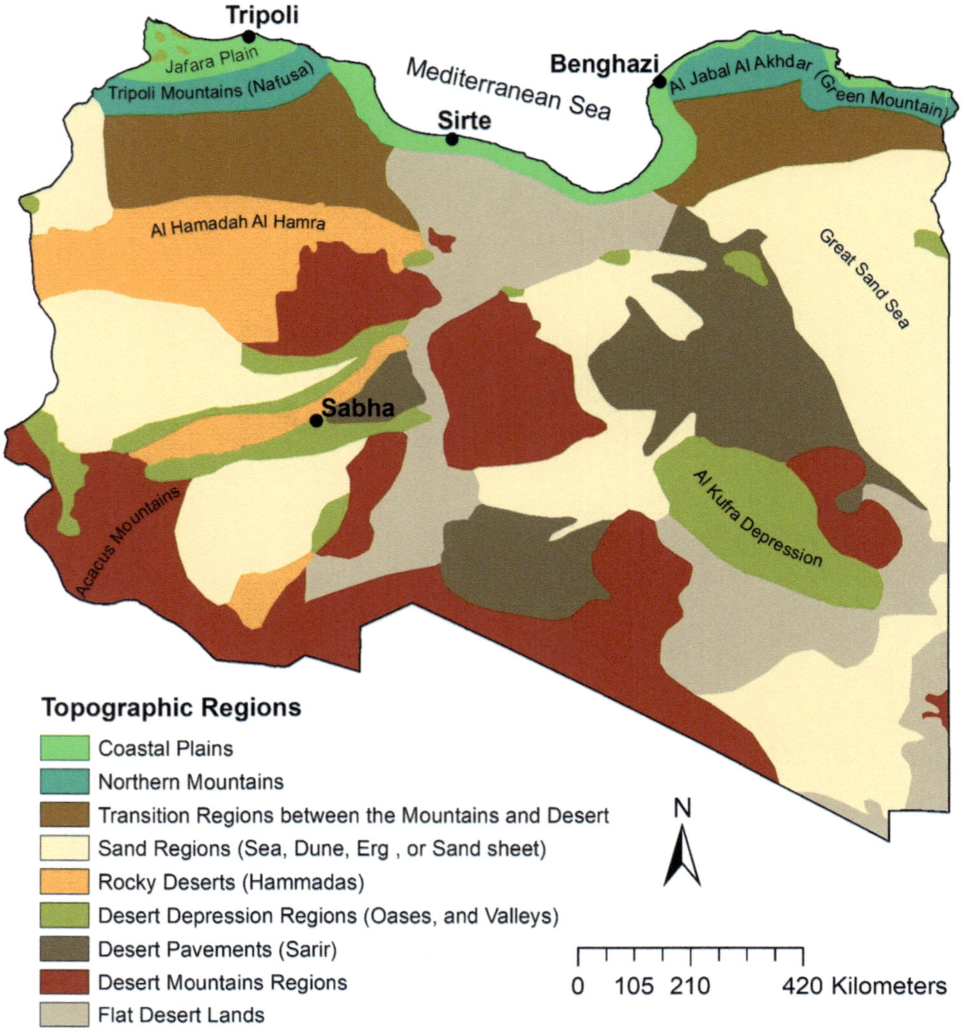

Fig. 3.7 Topographic regions of Libya. (Adapted from Ben Mahmoud 1995)

interspersed with some sand dunes of mixtures of the desert and marine origin, which is mostly rangeland with a low precipitation rate.

- **Desert region**: This region constitutes the greater part of the Libyan territory and consists of two parts:
  - **The eastern part**: This area is an extension of the eastern Libyan-Egyptian desert, and there are some oases (e.g., Al Jaghbub, Kufra, Jalu, Awjilah, and Tazirbu).
  - **The western part**: This area consists of a variety of surfaces, including Al Hamadah Al Hamra plateau in the north, Fezzan depression in the middle, and mountain heights in the south that gradually rise up to 2207 m above sea level in the Tibesti Mountains.

In light of this information, it is quite clear that the extent of the difference and the variation in the topography and the shapes of the land surface in Libya has a significant effect on the other soil forming factors. Thus, this explains the contrast in soil types across different regions in Libya.

3.2.4 Parent Materials (Geology)

The rocks (solid bedrock of earth's crust) from which mineral soils are formed are called parent materials. Rocks are broken down to parent materials (smaller rocks) and eventually become soil through the process of weathering. The effect of the parent material on the soil properties is obvious in the arid regions (i.e., Libya). The soil formation processes lead the soil to develop newly "acquired traits" that are soil properties that distinguish these soils from the parent materials (Ben Mahmoud 1995). In view of the variation in the parent materials of the Libyan soils, it is necessary to mention that the geological formations of the Libyan lands are wide-ranging in falling between Precambrian to Quaternary (UN Economic Commission for Africa-Addis Ababa1 1991). These geological layers are distributed over the basins of the groundwater, and some of them represent important aquifers (Fig. 3.8).

In terms of geology, Libya in general is a cratonic basin on the northern fringe of the African Shield. Precambrian

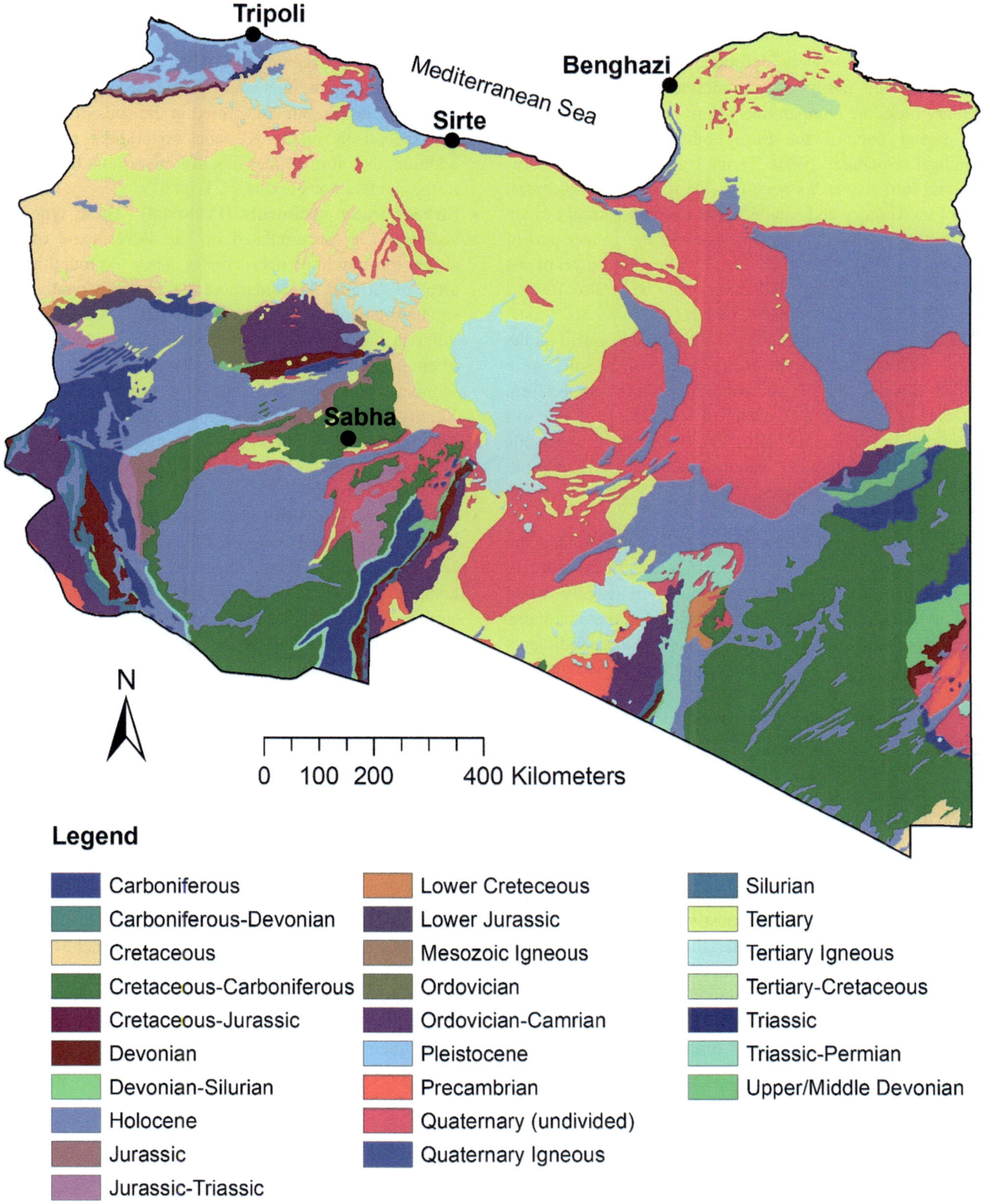

Fig. 3.8 Geological map of Libya. (Adapted from Persits et al. 2002)

rocks are found in the south and southeastern Libya and in northern Fezzan. By and large, Libya contains thick sequences of moderately deformed Paleozoic rocks, and aside from the northwest and northeast, Mesozoic sedimentary rocks are comparatively thin. Tertiary rocks occupy the greater part of the Sirte embayment and northern Benghazi (Goudarzi 1970). Tertiary and Quaternary extrusive and intrusive rocks occupy wide regions in the central part of the country and smaller areas in south-central Fezzan and northern Tripoli.Quaternary age and the greater part of the Libyan desert are both covered by immense gravel plains and sand dunes (these sand dune areas were also formed during the Quaternary time) (Goudarzi 1970).

In Libya, the soil mineralogy is strongly dependent on the parent material (or parent rock) from which soils were formed. Besides, dry climatic conditions affect both the form and rate of physical and chemical weathering of the soil parent material (Zurqani et al. 2019). Hence, the Libyan soils consist of the following parent materials (Solkhozprom Export 1980; Ben Mahmoud 1995):

- **Local (Residuum) parent material**: Local residuums the parent material that weathered in place from bedrock, typically millions of years old, and it is usually called (Eluvial), and these sediments are found in the mountains of Tripoli and the Al Jabal Al Akhda, and these sediments include Igneous rocks (Basalt) and Marine sedimentary rocks.
- **Transported parent material**: Transported parent material is material that is formed from the rocks as a result of weathering processes and then is transported to sites near or far from the parent rock by wind, water, and gravity. This type of parent material covers large areas in Libya and can be divided into
  - **Wind transported parent material (Eolian)**: Eolian is the sediments that were transported by the winds, and they are one of the most widespread Libyan soil materials. This parent material covers the areas of the Jafara Plain, the coastal areas, and some other regions such as Sarir, Kufra, Qaminis, Suluq, and Sirte. These sediments are in the size of grains of sand and thus are known as sandy sediments, or in the size of silt and are known as silty sediments, also called loess.
  - **Water transported parent material (Alluvium)**: Alluvium is the parent material transported by water, and it consists of igneous and sedimentary rocks. Within this type of parent material, there are several important types:
- **Floodplain sediments (Alluvial)**: These sediments cover the lands of the valleys and contain various amounts of sand, clay, and gravel, as well as some amounts of calcium carbonate and gypsum.
- **Deluvial (colluvial), or Mixed (Eluvial-Deluvial) sediments**: These sediments are found on talus drift and creep fans in the hilly regions in Al Jabal Al Akhdar and the Tripoli Mountains. Generally, these sediments are not formed as layers, but are formed in the soils of Vertisols and Inceptisols in the Al Jabal Al Akhdar region. In addition, they formed the soils of Entisols and Inceptisols in the mountainous region of Tripoli.
- **Stream water sediments (Proluvial)**: These types of sediments are accumulated on the lower parts of the slopes forming large debris cones. They are found in the areas of AlJabal AlAkhdar and the Tripoli Mountains in small waterways and some small valleys. Usually, they differ in the textures, are rich in stones, and stones are often intercalated with carbonates and gypsum. These sediments are considered to be the parent materials of the Xeralfs (Terra Rossa soils of the Mediterranean region). They may be also found combined with the sediments of the flood valleys, as is the case in the regions of Al Marj and east Gharyan.
- **Marine floor and Salt Lake (Sabkha) sediments**: These sediments are found along the Libyan coastal strip, which are shallow marine pools, consisting of layers of different textures. Usually, they contain saline or salt crust on the surface, and this type of parent material formed the Salids (one of accumulation of an excessive amount of salts that are more soluble than gypsum).
- **Coastal mound sediments**: These types of sediments can be distinguished based on their geological structure, which includes Qerqarish formation sediment (consists of sand, clay, and limestone deposits), GasrAl Hajj formation sediment (consists of pebbles that vary in size, number, and degree of cohesion associated with layers of caliche), and Jafara formation sediment (consists of sandy and silt deposits with layers of caliche rocks).

Due to this dissection and the effect of lithological variations, it is clear that the surface geological structure of the country varies from one region to another, and this variation is largely reflected in the parent materials of the Libyan soils. Hence, wherever these materials are found they differ in terms of their physical, mineral, and chemical properties, which in turn are nearly inherited to the soils in which they will be formed.

3.2.5 Time

Time is also a component for the other factors to interact with the soil. Over time, soils exhibit features that reflect the other forming factors (vegetation and climate act on parent material and topography). Soil formation is a continuous process and it generally takes several thousand years for significant changes to take place. Younger soils have some properties from their parent materials, but with age, the addition of organic matter, exposure to moisture, and other environmental factors, their features may be altered. The degree of aging depends on the intensity of other soil forming factors. Factors that slow soil formation include high lime content, high quartz content, or high clay content in parent material, hard rock parent material (resistant to weathering), low rainfall, low humidity, cold temperature, steep slopes, high water table, severe erosion, constant deposition, accumulations, and mixing by animals or man (Lamb and George 2018).

3.3 Processes of Soil Formation

In general, it is known that (mineral) soils form from the solid rocks of the earth's crust through a set of processes called soil formation processes. The process of soil formation is very complex and includes many overlapping processes. At any given geographical location, soil formation factors operate collectively to create ideal formation conditions. Controlled by the dominion and effectiveness of one or more soil formation processes, a particular type of soil is formed under these conditions (Ben Mahmoud 1995). As previously explained in Sect. 3.2 of this chapter, the factors of soil formation in Libya are mainly influenced by a hot and dry climate (low and irregular rainfall, high temperature, with a wide range of daily and seasonal temperature fluctuations, and strong wind speed). Thus, this climate dependency dictates particular types of soil formation processes that are more dominant in Libya and reduce, if not eliminate, the effectiveness of many of them.

3.3.1 Weathering Processes

Weathering processes are responsible for forming soil parent materials from the bedrocks. It is a group of processes that lead to changes in the natural state and chemical components of the rocks. It is basically a mixture of demolition and construction activities. With weathering, rock is disintegrated into smaller pieces (getting smaller and smaller) until they eventually reach the individual minerals from which they are formed.

3.3.1.1 Physical Weathering Processes

Physical weathering, also called Mechanical weathering, breaks rock into smaller pieces. These smaller pieces are just like the bigger rock, only differing in size. This means that the rock has changed physically without changing its chemical composition and crystalline form. The smaller pieces have the same minerals in just the same proportions as the original rock. These processes are among the most important processes that form the parent materials of the Libyan soils, especially in the western, central, and southern regions. The following are the most important physical weathering processes in Libya (Ben Mahmoud 1995):

- **Daily and seasonal temperature fluctuations**: The fluctuations of the daily and seasonal temperature lead to inconsistent fragmentation of rocks into uneven large portions, due to the difference in the thermal expansion coefficients of its minerals. This process is more effective in the southern desert regions (desert areas).
- **Erosion**: Erosion is the action of surface processes (such as water flow or wind) that removes soil, rock, or dissolved material from one location on the Earth's crust and then transports it to another location (Sriwati et al. 2018). In Libya, wind deposits (sand and silt) are considered one of the most important parent materials of the Libyan soils that commonly widespread across the country.
- **The mechanical action of sea waves**: This process is responsible for breaking up the rocks on the seabed. When the waves are high, sea waves throw these products on the beaches, forming marine parent materials or coastal sand.

3.3.1.2 Chemical Weathering Processes

Chemical weathering processes are considerably more varied and complex. They include simple solution, hydrolysis, hydration, and oxidation-reduction. These processes involve the alteration of the chemical and mineralogical composition of the weathered material which leads to the formation of new minerals. Water is the main determining factor for these processes, wherein its absence, the other essential factors for chemical reactions such as heat, gases, and microorganisms on their own seem incapable of playing any role. Therefore, the chemical weathering processes are more active in the regions that are exposed to a sufficient amount of rainfall. Thus, the impact of chemical weathering processes during the formation of the parent material in Libya is limited to areas that receive precipitation rates of more than 200 mm/year. These areas are located in Al Jabal Al Akhdar, the Tripoli Mountains, and some coastal areas (Ben Mahmoud 1995).

3.4 Soil Forming Processes

Soil forming processes are reflected in the different horizons developed within the soil profile. These processes can be limited to two main groups: specific processes of soil formation, and composite processes of soil formation. The following is a review of some of these processes that are specifically predominant in Libya (Ben Mahmoud 1995):

3.4.1 Specific Processes of Soil Formation

3.4.1.1 The Processes of Adding Mineral and Organic Materials to the Soil Surface (Additions)

- **Accumulation of organic material on the soil surface (Littering)**: This process is responsible for the development of the surface horizon (A1). The characteristics of this horizon differ from one region to another according to the type and density of vegetation. Taking into consideration that the vegetation cover in Libya is typically sparse or absent, the amount of accumulated organic material in the Libyan soils is considered very small. However, with the rise in temperature, the decomposition rates of the organic matter will be very high. Thus, an Ochric horizon is formed. This horizon is the most widespread soil surface horizon in the Libyan soils. In the other regions of the country (i.e., Al Jabal Al Akhdar) where the distribution of the vegetation cover is relatively high, lower temperatures lead to the formation of the organic-mineral horizon called Mollic, which is characteristic of Rendolls.
- **Mineral accumulation processes (Cumulization)**: These processes include both the accumulation of soil material by wind and water. In Libya, these processes are mainly responsible for the formation of sand dunes and Psamments (which is most widespread in the Libyan soil). As well as the Fluvents which are formed along flood plains by alluvial (water-deposited) soils. Also, these processes lead to a delay in the development of the soil profile, due to the lack of sufficient time for other soil formation processes to involve in the development of the soil profile.

3.4.1.2 The Processes of Loss Mineral and Organic Materials from the Soil Surface (Losses)

- **Leaching (removal of soluble materials)**: Leaching is the removal of soluble materials from one zone in soil to another via water movement in the profile. These processes include washing the water-soluble salts (e.g., calcium and gypsum) inside the soil profile to a deep depth. The occurrence of this process requires an adequate amount of water. Therefore, the complete washing of these materials from the soil profile is limited in areas that receive high precipitation rates such as the northeastern region of the country (i.e., Al Jabal Al Akhdar).
- **Erosion (removal of the surface layer)**: The occurrence of this process leads to the loss of the soil surface layer through soil erosion by wind and/or water. These processes are very active in Libyan soils and lead to a delay in the development of soil profile. As a result, a (Truncated profile) is formed, and that is what usually happens in the soils of Xeralfs in the Al Jabal Al Akhdar region.

3.4.1.3 Transfers of Organic and Mineral Materials Within the Soil Profile (Transformations)

These processes are similar to the previous processes, but the materials are not completely lost from the soil profile, rather they are transported from one place and deposited in another place within the same soil profile. These processes also need adequate water, but not the same amount of water needed for the leaching processes. Such processes are responsible for the formation of many subsurface diagnostic horizons, and among these processes are the following:

- **The process of transporting the water-soluble salts from the surface by rainwater**: Salts are deposited in the subsurface layers forming the saline horizon (Salic), in which the percentage of salts is more than 2% (i.e., Salids).
- **The process of transporting the calcium carbonate from the surface by rainwater**: During this process, calcium carbonate is deposited at varying depths according to the amount of precipitation. This process is responsible for the formation of the Calcic horizon, the Petrocalcic horizon, and the Caliche layer, as in the case of Calcid soils.
- **The process of transporting the gypsum (dihydrate calcium sulfate) from the surface by rainwater**: In this process, gypsum is deposited at different depths, according to the precipitation rates in the area. This process is responsible for the formation of the gypsum horizons (Gypsic or Petrogypsic). These horizons may be formed by transferring gypsum from the groundwater level close to the surface, as in some Sabkhas. Usually, gypsum in this case is below the saline horizon (Salic), as in the case of the Gypsids.
- **The process of transporting the Silicate clay (Illuviation)**: The process of transporting silicate clay minerals to the subsurface layers is a water-assisted transport in a basically vertical direction. First, the hydrolysis process needs to take place to form secondary clay minerals in the soil. Thus, this process requires a sufficient amount of water (high humidity conditions). Therefore, this process is very rare and occurs in Libyan soils. However, it was found that the Libyan soils contain the clay horizon (Argillic) and the Sodic horizon called Natric. Scientists have indicated that the existence of the clay horizons in arid soils can be explained by one of the following reasons (Dregne 2011; Ben Mahmoud 1995):

- The formation of these horizons is due to the presence of clay materials transferred by the mechanical migration method (Lessivage).
- The clay horizon may be developed under different climatic conditions (wetter) than it is now.
- The clay in the subsurface layers do not move from the surface but rather are formed in the same place.

- **The process of biological or physical mixing of soil materials (Pedoturbation)**: In this process, the surface horizons are mixed with the subsurface horizons, through the transfer of the soil through thick and deep cracks. As in the case of Xererts in some regions of Al Jabal Al Akhdar.
- **The process of formation of a new mineral or organic species (Synthesis)**: These processes require an abundant amount of water. Thus, they are almost absent or non-existent in the Libyan soils, especially in arid regions (i.e., southern), and among these processes are the following:
  - **The process of breaking down the mineral and organic materials (Decomposition)**: Many soil living organisms are responsible for this process. This process results in intermediate compounds that are not completely degraded, known as Humus (Organic Colloids). Hence, it plays a crucial role in improving soil properties. However, the production of humus in Libyan soils is very small due to the high temperature and good aeration conditions.
  - **The process of hydrolysis and formation of silicate clay minerals**:
  - During this process, the minerals in the rocks are chemically destroyed. Thus, new minerals (clay minerals) are formed. Likewise, water must be available for this process to occur. Accordingly, the formation process of silicate clay minerals in Libyan soils is very little or no process. Given the fact that the availability of these minerals in Libyan soils, it can be suggested that is due to their presence in the parent material (inherited traits) (Ben Mahmoud 1995).
- **Reactions that change the chemical composition of soil mineral matter**: The occurrence of these processes depends on the availability of water and aeration. Thus, it can occur under special circumstances in Libyan soils, and it includes many processes, namely (Ben Mahmoud 1995; Naguib and Khidr 1989)
  - **Hydration**: It is the chemical combination of the polar water molecules with another substance. The water molecules do not dissociate, but attach as a shell to the other substance. For example, the transformation of Hematite (red) into Limonite (yellow):

$$2Fe_2O_3 + 3H_2O - - - - - \rightarrow 2Fe_2O_3.3H_2O$$

$$\text{Hematite (red)} - - - - - \rightarrow \text{Limonite (yellow)}$$

The formation of limonite from hematite is a typical hydration reaction in soils and one that causes soil color to change.

- **Carbonation**: It is the dissolution of carbon dioxide in water. First, carbonic acid H_2CO_3 is formed by the reaction of CO_2 with H_2O and released into the soil solution by root or microbial respiration. Then, basic oxides are converted to carbonates through a reaction with carbonic acid, which has an effective role in dissolving some minerals in rocks, such as calcium and magnesium carbonates, to more water-soluble forms:

$$CaO_2 + H_2O - - - - - \rightarrow H_2CaO_3$$

$$CaCO_3 + H_2CO_3 - - - - - \rightarrow Ca(HCO_3)_2$$

$$\text{Calcium carbonate} + \text{CarbonicAcid} - - - - - \rightarrow \text{Calcium bicarbonate}$$

- **Oxidation-Reduction**: It is a set of coincident reactions involving electron transfer between atoms. This process affects the rocks and minerals containing the oxidized or reduced atom, and in turn, the compound's susceptibility to physical or chemical weathering. This process leads to chemical changes that help break down rocks and minerals. For example, the oxidation of the minerals containing Ferrous iron to Ferric iron, under proper conditions of water and air:

$$Fe_2O_3 \overrightarrow{-O_2(\text{reduction})} 2FeO \overrightarrow{+O_2(\text{reduction})} Fe_2O_3$$

$$\text{Ferric oxide (red)} \rightarrow \text{Ferrous oxide(black)} \rightarrow \text{Ferric oxide (red)}$$

Ferrous iron is highly soluble and mobile in the soil profile, but exposure to O_2 easily oxidizes it to ferric iron. Ferric oxide is the tan, brown or red concretions, and particle coatings can be visibly observed in the soil.

- **The microbiological processes of soil microorganisms (Decomposition)**: These processes occur by particular types of bacteria at the sites where temperature and moisture provide conditions that are conducive for the multiplication of microorganisms. These bacteria perform some transformations and form new chemical compounds, and among these processes are the following:

- Weathering of soil minerals and release of some important elements such as potassium, phosphorous, iron and, sulfur.
- The biological atmospheric nitrogen fixation.
- Mineralization (release of mineral constituents from organic matter).
- Nitrification (the oxidation of the nitrite to nitrate).
- Denitrification (the reduction of nitrates or nitrites that often results in the escape of nitrogen into the air).
- Process of reverse ionization occurs, i.e., converting metallic nitrogen into a gaseous form.
- The oxidation of iron and other compounds.

All these processes and others may take place in the Libyan soils once the conducive conditions are available.

3.4.2 Composite Processes of Soil Formation

Composite processes are a group of specific soil formation processes that work together to form a particular type of soil, under certain conditions of soil formation factors. This does not mean that these soil formation processes work simultaneously together but may depend on each other. Usually, each group of these processes is called by the name of the soil formation process that produces it. There are many processes that contribute to the formation of the Libyan soils, the most important of which are the following (Ben Mahmoud 1995):

- **Salinization (accumulation of soluble salts)**: Insufficient sedimentation for salt accumulation due to high temperatures and evaporation leads to the accumulation of salt in the soil profile (Zurqani et al. 2019). The salinization process redistributes the water-soluble salts within the soil profile. This process takes place along the Libyan coastal strip, in desperation areas with poor drainage, or the dry marine pools in the contiguous areas of Benghazi. Salinization may occur naturally or because of conditions resulting from management practices. The soils formed by this process are called Salids or Solonchack. These soils are characterized by the fact that the percentage of soluble salts may reach 2%, as well as the presence of the saline horizon. These soils sometimes form what is known as the white salt crust, and the process of redistribution of the water-soluble salts within the soil profile in two ways is as following:
  - Transfer of salts from the surface layers by the washing processes and deposit in the subsurface layers.
  - Transfer of salts from bottom to top when the groundwater is close to the surface.

The preferred conditions to activate this process are a hot and dry climate, low elevation, the close groundwater to the surface, and the presence of the parent material that contains high soluble salts.

- **Salonization (sodicity)**: Sodicity refers to high sodium (Na) levels in the soil. Soil sodicity is measured either through its Exchangeable Sodium Percentage (ESP) or Sodium Adsorption Ratio (SAR) (Zurqani et al. 2018). During the Salonization process, the sodium element replaces other chemical elements on the absorption complex of silicate clay minerals and humus. Also, a combination of sodium with carbonates occurs in the soil solution to form sodium carbonate (Na_2CO_3). Thus, the presence of large amounts of sodium in the soils can cause soil dispersion, which prevents the formation of soil aggregates. Sodium weakens the bonds between soil particles, which leads to disperse and disburse of these particles. This process is responsible for the formation of the Natric horizon, which is characterized by a prismatic or vertical structure that limits the extension of roots in the soil. Consequently, the sodic soils are formed (Natriargids or Solonetz).

The favored conditions for this process to occur are the hot, dry climate, inability to completely wash out the salts from the soil profile, and the presence of the parent material with clayey texture. The predominant salt in this parent material is sodium. This process takes place in limited and very small areas in Libya (i.e., Al Jabal Al Akhdar region and "Benghazi plain"), where the soils have a clayey texture. Some of these soils can be managed by leaching out the sodium present in the soil through the process called Desolonization. It is the process that is used to reclaim the sodic soils by adding agricultural gypsum (lime) to replace sodium with calcium, with the addition of irrigation water to wash salts from the soil profile.

- **Calcification (Liming process)**: This is the general process by which calcium carbonate, and sometimes calcium and magnesium carbonate, accumulates in the soils. Usually, the accumulation of calcium carbonate in subsoil horizons is coupled with its removal from overlying horizons, linked by the downward translocation of the soluble ions that precipitate to form calcium carbonate ($CaCO_3$). This is according to the rates and conditions of washing. This process is responsible for the formation of calcareous soils, including the formation of the Calcic and Petrocalcic horizons. In order for this process to occur, calcium carbonates (and sometimes calcium and magnesium carbonates) are transported or redistributed by rainwater or the groundwater close to the surface (Dregne 2011).

- **Ferrisiallitization**: Ferrisiallitization is the process that is responsible for the formation of soil Xeralfs or Ferrosiallitic, which is located in Al Jabal Al Akhdar and the Benghazi Plain. In this process, the soluble salts and the base elements are partially washed from the upper layers of the soil profile. As a consequence, new silicate minerals are formed and accumulated in the place of their formation simultaneously with the accumulation of iron oxides and hydroxides. In drought periods, the distinctive red color dominates the color of these soils, as a sign of Dehydration. This process may be associated with the other specific composite soil forming processes, such as the process of hydration, salinization, and others, which may cause differences in soil properties from one region to another (Selkhozprom Export, 1980).

3.5 Conclusion

Generally, aside from Al Jabal Al Akhdar and a portion of Tripoli Mountains (Jabal Nafusa), Libya is influenced by an arid and semi-arid climate where vegetation cover is usually sparse or absent. Soil forming factors and the formation processes of diagnostic horizons in different regions of Libya are mainly influenced by these characteristics. Consequently, the development of soil profiles is extremely limited, which results in undeveloped or partially developed soils, such as Entisols and Aridisols, due to the ineffectiveness of organic matter accumulation and washing processes (additions, losses, transformations, etc.). Therefore, the profiles of these soils are often devoid of developed subsurface diagnostic horizons. However, in somewhat conducive local conditions, Calcic, Salic, or Gypsic horizons may be formed. The situation is slightly different in coastal areas, especially in the Al Jabal Al Akhdar and Tripoli Mountains regions, where precipitation rates are slightly higher. This climate condition encourages the presence of vegetation cover, which activates some of the formation processes in these regions. This includes the accumulation of organic matter on the surface, or the formation of silicate clay minerals, moving it and accumulating it within the soil profile, and developing the Mollic, Cambic, or Argillic horizons forming the most relatively developed soils in the country. Most of the parent material of Libyan soils contains a high percentage of calcium carbonate and exists on a steep topography, which disrupts the functioning of the climate and vegetation and consequently the soil developments. It should also be noted that some Libyan soils, especially those in the south (in the desert), are not related in their traits and characteristics to the current climate. Rather, their formation is due to the pluvial period, which can be traced back to the Quaternary and Early Modern eras.

Acknowledgements Professor Khaled Ben Mahmoud would like to thank and express his appreciation and gratitude to the late Prof. Dr. Robert D. Heil, Professor of Soil Genesis, Classification and Mapping at Colorado State University, Fort Collins, Colorado, USA. The late Prof. Heil was the Academic Supervisor for Prof. Ben Mahmoud in his MSc and Ph.D. (1971–1977), from which he learned, besides the academic and practical aspects, seriousness in work, humility, participation by others, and encouraging them to work in a team spirit. This is the same policy that Prof. Ben Mahmoud followed with his colleagues who graduated during the past years from Libyan universities. Professor Ben Mahmoud's contribution to this chapter is dedicated to the spirit of Prof. Robert D. Hiel.

References

Al Hajjaji S (1989) The New Libya (a geographical, socio-economic and political study), Tripoli Publishing Center. Tripoli, Libya

Ben Mahmoud KR (1995) Libyan soils (Their Genesis, Classification, Properties and Agricultural potentials) NASR, Tripoli, Libya, pp 615 (in Arabic)

Ben Mahmoud KR (2013) Towards a national strategy for the sustainability of natural resources and enhancing food security in Libya, pp 123 (in Arabic)

Dregne HE (2011) Soils of arid regions. Development in Soil Science (Vol. 6). Elsevier Scientific Publishing Company, Amsterdam, Netherlands, pp 236

Elhawej AZ, Elaalem MM (2012) A guide morphological description and classification soil profile in the field. pp 80 (in Arabic)

Fick SE, Hijmans RJ (2017) WorldClim 2: new 1 km spatial resolution climate surfaces for global land areas. Int J Climatol 37(12):4302–4315

Goudarzi GH (1970). Geology and mineral resources of Libya-a reconnaissance (No. 660). US Govt. Print. Off.,. https://pubs.er.usgs.gov/publication/pp660. Accessed 25 October 2020

Government of Libya, Secretariat of Planning, Surveying Department (1978) The National Atlas of Libya. Tripoli, Libya

Lamb J , George WR (2018) Five factors of soil formation. https://extension.umn.edu/soil-management-and-health/five-factors-soil-formation. Accessed 25 October 2020

LIB/00/004. 2009. Mapping of Natural Resources for Agriculture Use and Planning in Libya Project, Atlas of Natural Resources for Agricultural Use in Libya. http://www.fao.org/geospatial/resources/detail/en/c/1024739/. Accessed 11 April 2019

Naguib MM, Khidr M (1989) Foundations of pedology. Modern Arabic office for printing and publishing, Alexandria Egypt

Persits F, Ahlbrandt T, Tuttle M, Charpentier R, Brownfield M, Takahashi K (2002) Map showing geology, oil and gas fields and geologic provinces of Africa, Ver 2.0. USGS Open File report 97-470 A

Sharaf AA (1971) Libya geography. Al MaarifFacility in Alexandria, Egypt

Solkhozprom Export. Soil Ecological Expedition. USSR (1980) Soil studies in western zone, eastern zone, and central zone of Libya. Ministry of Agri. and Land Development. Tripoli. Three separate volumes. (in English)

Sriwati M, Pallu S, Selintung M, Lopa R (2018) Bioengineering technology to control river soil erosion using vetiver (VetiveriaZizaniodes). E&ES 140(1):012040

UN Economic Commission for Africa-Addis Ababa (1991) Study on large-scale water transfer in Africa. United Nations Economic Commission for Africa (ECA) https://www.ircwash.org/resources/study-large-scale-water-transfer-africa. Accessed 25 October 2020

Zurqani H, Mikhailova E, Post C, Schlautman M, Sharp J (2018) Predicting the classes and distribution of salt-affected soils in north-west Libya. Commun Soil Sci Plant Anal 49(6):689–700

Zurqani HA, Mikhailova EA, Post CJ, Schlautman MA, Elhawej AR (2019) A review of libyan soil databases for use within an ecosystem services framework. Land 8(5):82

Dr. Khaled R. Ben Mahmoud is a senior professor of soil pedology. He is a distinguished soil scientist with wide range of scientific and working experiences in Libya and abroad. He received his Ph.D. in 1977 from Colorado State University (USA). His major teaching activities and research are at the Department of Soil and Water, Faculty of Agriculture, University of Tripoli, Libya. He has been teaching undergraduate and graduate students in Libya and has supervised many M.Sc. and Ph.D. thesis and dissertation in Libya and abroad. He focused mainly on soil resources, land evaluation, and mapping. He contributes to soil science in Libya and abroad as the Head of the Department of Soil and Water Sciences, Faculty of Agriculture, University of Tripoli, Tripoli, Libya (1978–1983), Dean of the Faculty of Agriculture, University of Tripoli, Tripoli, Libya (1983–1986), Chairman of the Committee for Libyan Agricultural Mapping, Tripoli, Libya (1988–1990). A Research Coordinator, University of Tripoli, Tripoli, Libya (1997–2001), Member of the Scientific Council of Agricultural Research Center (ARC), Tripoli, Libya (2000–2002), Member of the Scientific Council of the College of Higher Studies at the University of Tripoli, Tripoli, Libya (2000–2002), Member of the Scientific Committee of Libya Centre for Remote Sensing and Space Sciences, Tripoli, Libya (1998–2006), A National Coordinator of the Project implemented by FAO: Mapping of Natural Resources for Agricultural Use and Planning Tripoli, Libya (2000–2006), A General Coordinator of the National Programme for the Development of Vegetation Cover, Tripoli, Libya (2004–2006). Director of the Division of Land and Water Uses, The Arab Centre of the Studies of Arid Zones Lands (ACSAD), The League of Arab States, Damascus, Syria (2007–2011), Head of the Programme: Monitoring and Combating of Desertification in the Arab World, The Arab Centre of the Studies of Arid Zones Lands (ACSAD), The League of Arab States, Damascus, Syria (2007–2011), Director General of Eddayra Co. (Agricultural and Environmental Consulting and Services), Tripoli, Libya (2011–to date). Engaged as a local and international expert to control, evaluate, approve, and accept of many soils studies and mapping which have been implemented by different foreign and local companies, agencies, and offices in different parts of Libya and abroad (1977–to date). More than 100 country reports, research and scientific papers, and other publications were published in different local and international scientific journals and symposiums (in Arabic and English). He is the author of more than 10 books that have been written in Arabic. The most important ones are: Sandy soils of Libya (1981), The study of soils in the field (1984), Principles of Soil Science for Agricultural Secondary and Technical Schools (1985), Fundamentals of Soil Science and its Relation to Plant Growth (1995), Libyan Soils (Their Genesis, Classification, Properties, and Agricultural Potentiality (1995), Fundamentals of Soil and water Science (2001), Issues on Desertification and Land Degradation in the Arab Region (2010), Towards a National Strategy for the Sustainability of Natural Resources and Enhancing Food Security in Libya (2013), and The Libyan Desert (its natural potentials and prospects for investment) in press. Finally, he was appointed as a technical advisor to the President of the Libyan Presidential Council for Agriculture, Tripoli, Libya (2018–to date).

Dr. Hamdi A. Zurqani is one of the faculty members at the Department of Soil and Water Sciences, Faculty of Agriculture, University of Tripoli, Tripoli, Libya. Dr. Zurqani is a recognized expert as a result of his internationally acclaimed work in the areas of environmental information science, remote sensing, land evaluation, sustainability, pedology, and soil science education. He has conducted research across the world, including the United States of America and Africa. Dr. Zurqani is a distinguished soil scientist with a wide range of scientific and working experiences in Libya and abroad. He received his M.Sc. (2010) from the University of Tripoli, Tripoli, Libya and Ph.D. (2019) from Clemson University, Clemson, SC, USA. His major research and teaching activities at the University of Tripoli have focused mainly on soil genesis and classification and the environmental information sciences (Remote Sensing and Geographic Information System). He has published broadly in many journals (e.g., Nature "Scientific Reports", Geoderma; International Journal of Applied Earth Observation and Geoinformation; Journal of Geological Sciences; Land; Frontiers in Environmental Science; Communications in Soil Science and Plant Analysis; and others). Dr. Zurqani is a member of the Editorial Board for Remote Sensing (MDPI) Journal, counseling outcome and research evaluation. He also was appointed to serve as a Guest Editor for the Special Issue "Applications of Remote Sensing in Earth Observation and Geo-Information Science". In addition, Dr. Zurqani conducted peer-review for many journals including Journal of Environmental Informatics, Applied Sciences, SN Applied Sciences, Remote Sensing, Heliyon, Geosciences, Land, Water, Agronomy, Agriculture, Sustainability, Arid Land Research and Management, International Journal of Environmental Research and Public Health, Natural Hazards, and Conference of the Arabian Journal of Geosciences. He is also one of the authors of the lab manual entitled, "GIS Exercises for Natural Resource Management". Dr. Zurqani has been the recipient of numerous awards and honors: Recipient of Douglas R. Phillips Award for Graduate Students, Department of Forestry and Environmental Conservation, Clemson University, April 12, 2019; the First Place Best Judged Poster (CAFLS) at the GRADS 2019: Clemson Student Research Forum on April 4, 2019; the Second Place Poster at the 11th Clemson Biological Sciences Annual Student Symposium, April 6, 2019; the Second Place Best Judged Poster at the Clemson Student Research Forum on April 4, 2018; and the Third Place Poster at the 9th Clemson Biological Sciences Annual Student Symposium, February 25, 2017. Dr. Zurqani conducts cutting-edge research in the field of environmental information science, remote sensing, land use management/planning, change detection of landscape degradation, and geographic information system (GIS) models. He has focused on his research efforts on the development of new technologies in the field of environmental information sciences, geo-intelligence (advanced geo-information science and earth observation, machine and deep learning, and big data analytics), remote sensing, land evaluation, pedology, land use management/planning, monitoring and evaluating sustainable land management, change detection of landscape degradation, and geographic information system models.

4 Soil Classification and Properties

Mukhtar M. Elaalem, Hamdi A. Zurqani, Khaled R. Ben Mahmoud, and Azzeddin R. Elhawej

Abstract

As mentioned in Chap. 3, soils and their properties in Libya are influenced to the great extent by nature and conditions in which these soils were formed. Hence, the types of soil formed in Libya differ from one place to another, and this difference is related to the diversity in the factors responsible for soil formation. In this chapter, we discussed the most common Soil Classification Units found in the country and reviewed the main properties of the Libyan soils. As mentioned also in Chap. 2, Many soil studies have been carried out in Libya over the last four decades. These soil survey reports differ in their contents, types of maps, the scale of mapping, classification systems, methods of soil analysis, and the criteria on which the interpretation of data is based. The major soil classification systems used in Libya are the Russian soil classification system, the U.S.A Soil Taxonomy, and the FAO soil classification system. In the early 1990s, Libyan soil experts decided to adopt the U.S. A. Soil Taxonomy for future soil survey work and converted the names of soil taxonomic units from the Russian soil classification system to the U.S.A. one. The taxonomic units for the Libyan soils were discussed according to the Russian and U.S.A soil classification systems. According to the U.S.A. Soil Taxonomy, Soil Taxonomic Units of Libya are either Entisols or Aridisols. Alfisols, Mollisols, and Vertisols can only be found in Al Jabal Al Akhdar. Inceptisols are placed in both Al Jabal Al Akhdar and Jabal Nafusa. Soil Properties of the main soil taxonomic units at the great group levels were also discussed. The major soil limiting factors that are risking the agricultural activities in the country were also defined and will be discussed further in the next chapter.

Keywords

Soil formation • Mediterranean climate • Soil genesis and classification • Soil mapping • Soil characteristics • Agricultural capability

4.1 Introduction

As discussed in Chap. 3, the formation and genesis of soils (at any region in the world) are quite complex, as they contain the amassing (adding), loosing, transporting, and generating processes. As it is widely known, these processes take place under the effect of factors responsible for soil formation. The major five factors responsible for soil formation in Libyan soils (like any soils formed in other regions of the world) are climate, biota (mainly vegetation cover), parent materials, topography, and time. In 2013, Ben Mahmoud stated that the soil types formed in Libya vary from one place to another and this variation is related to the diversity found in the factors responsible for soil formation.

It is important here to mention that, by eliminating the coastline, north mountain areas (in the east and west regions), and some scattered depressions and old wadis in southern Libya, which contain relatively developed soils (as a scientific term), the majority of Libyan lands are covered with non-soil covers. According to Ben Mahmoud (2013), this is the case for about 90% of total Libyan lands including Sands and Sand Dunes, Rocky Deserts (Hammadas), Desert Pavements (Sarrier), and Saline Depressions (Sebkha). The reset of the lands (about 3 million hectares in the northern

M. M. Elaalem · H. A. Zurqani (✉) · K. R. Ben Mahmoud · A. R. Elhawej
Faculty of Agriculture, Department of Soil and Water Sciences, University of Tripoli, Tripoli, 13538, Libya
e-mail: h.zurqani@uot.edu.ly

M. M. Elaalem
e-mail: m.elaalem@uot.edu.ly

K. R. Ben Mahmoud
e-mail: benmahmoud.kh@gmail.com

A. R. Elhawej
e-mail: azzeddin47@gmail.com

H. A. Zurqani (ed.), *The Soils of Libya*, World Soils Book Series,
https://doi.org/10.1007/978-3-030-66368-1_4

part of Libya) which receives precipitation of more than 200 mm/year and exposed to the Mediterranean and semi-Mediterranean climates, cover with soils which are varied in their development according to their soil-forming factors and processes that exposed to (see Chap. 3). The difference in soils in these areas actually is due to the combined effect of soil-forming factors which determined the activity of specific soil forming process (s), for example, the Al Jabal Al Akhdar region is semiarid to subhumid (annual precipitation of 300–600 mm/year), vegetation cover is forest trees or shrubs, parent materials are calcareous sedimentary rocks, and the terrain is mountainous. Under these conditions, formation processes called fersiallitization are activated accompanied by specific soil formation processes such as illuviation, dehydration, and calcification. Consequently, a diagnostic subsurface horizon (Argillic) of red clay soils is formed (the soil type is called Xeralfs) (Solkhozprom Export 1980; Ben Mahmoud 1995).

As for the northern areas of western Libya, specifically in the Jafara Plain, where the parent materials are sandy sediments, the climate is semiarid to arid (annual precipitation of 200–350 mm/year), and the semidesert vegetation is weakly widespread, the activity of many processes of soil formation are limited. Therefore, a soil of limited development was formed which has no diagnostic horizons except for the surface diagnostic horizon (Ochric), and thus the soil type spread in these areas are Psamments (Ben Mahmoud 1995). On the other hand, in the semidesert and desert areas in southern Libya, where there is a desert climate (annual precipitation of less than 50 mm/year) and an absence or scarcity of vegetation cover, soils are only formed in the desert depressions or old valleys where the water table level is fairly close to the land surface. Under these conditions, a number of specific soil formation processes such as Salinization, Solonization, and calcification may occur, forming only Entisols or Aridisols (Ben Mahmoud 1995).

4.2 Soil Classification Systems Used in Libya

4.2.1 The Russian Soil Classification System

The first soil pedological classification system used in Libya is the Russian soil classification system (Selkhozpromexport 1980). Soils of 3 million hectares in the northwestern region and northeastern regions (areas receiving annual precipitation of more than 200 mm/year) were investigated using the Russian soil classification system. This study was conducted by the Soil-Ecological Expedition of v/o "Selkhozpromexport", the Agricultural Research Centre (ARC), University of Tripoli, and the Ministry of Agriculture. The taxonomy of the Soviet soil pedology system was used for the elaboration of the soil classification, and the soil nomenclature generally applied to characterize the soil mantle of the Mediterranean countries was also partially used. Classes and subclasses have been singled out on the basis of the classification structure for the tropics and subtropics given by Zoon (1974). The definitions of the Russian terminology system used are summarized below:

- **Class**: A class unites soils of similar mineral part composition, the similarity being caused by the nature and direction of soil formation, as well as by peculiarities of origin and age of parent materials (weathering crusts).
- **Subclass**: A subclass unites soil types with similar combinations of the conditions of their formation connected with the development processes which are conditioned by the composition and properties of the soil-forming rock, as well as peculiarities of climatic regimes.
- **Type**: A type unites soils which develop under similar (typical) biological, climatic, and hydrological conditions, and which have a similar soil profile structure, and generally, similar properties. Soils of a single type are characterized by common origin, migration, transformation, and accumulation of substances. Their genesis is connected with a distinct manifestation of the soil formation processes, including potential combinations with other processes.
- **Subtype**: A subtype embraces soils within a type, varying in quality as far as the intensity of manifestation of the main and secondary elementary processes of soil formation is concerned. Subtypes represent stages of an evolutionary transition of one type into another. While reflecting the peculiarities of soil development, subtypes preserve a general typical structure of the profile, but, at the same time, possess some specific features of their own.
- **Genera**: A genus includes soil groups within a subtype. A genus reflects soil properties connected with the influence of local factors, manifestation of the features caused by a peculiar character of parent material influence, and chemical composition of groundwater. The given classification distinguishes soils into genera according to a combination of their calcareousness, leachedness, solonetzicity, and salinity, as well as to the combination of these properties. Table 4.1. and Fig. 4.1 shows the scheme of soil division into classes, types, subtypes, and genera in northwestern region, while Table 4.2 and Fig. 4.2 shows classes, types, subtypes in northeastern region.

Table 4.1 The scheme of soil division into classes, types, subtypes, and genera in northwestern region

Soil type	Soil subtype	Soil genus
Siallitic cinnamon	Typical	Carbonate, carbonate saline, leached
	Crust	
Reddish-brown arid	Differentiated	Carbonate, carbonate saline, carbonate solonetzic-saline and carbonate gypsic
	Differentiated crust	Carbonate and carbonate saline
	Slightly differentiated	Carbonate, carbonate saline, carbonate solonetzic-saline, carbonate gypsic, and leached
Reddish-brown arid	Slightly differentiated crust	Carbonate, carbonate saline, carbonate gypsic, and leached
	Non- differentiated	Carbonate, carbonate saline, leached, and non-carbonate
	Non-differentiated crust	Carbonate, carbonate saline, and leached
Alluvial	Slightly differentiated	Carbonate
Lithosols	Cinnamonic	Carbonate, carbonate saline, and carbonate gypsic
	Reddish-brown	
Crusts	Non-monolithic	
Solonchaks	Hydromorphic, Hydromorphic crust, and Hydromorphic Sebkha	

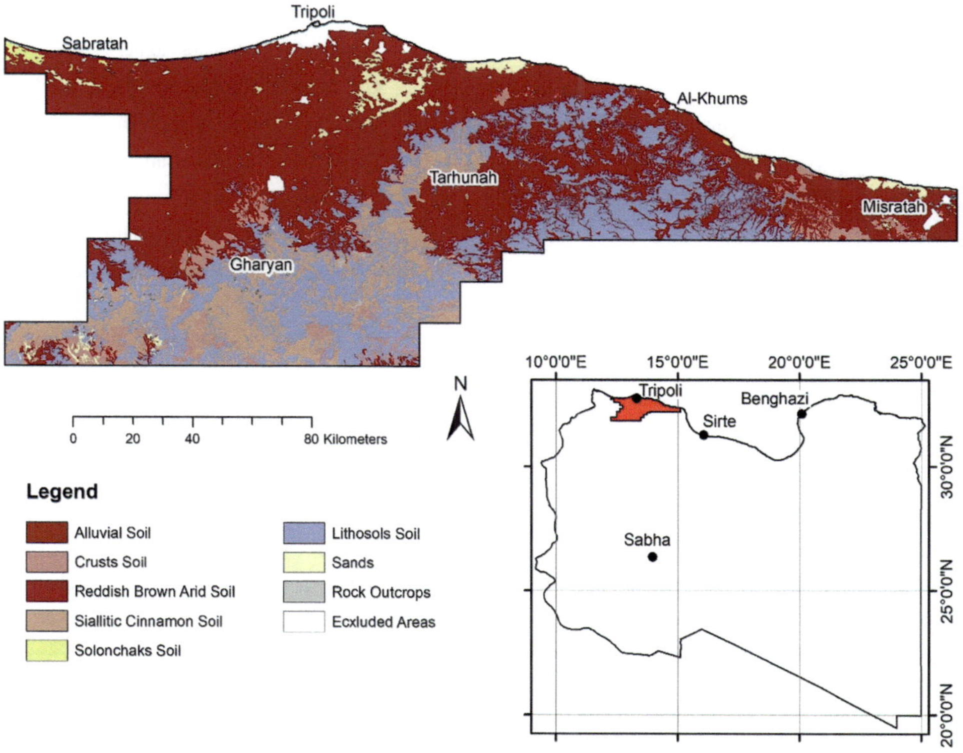

Fig. 4.1 Soil Types in north-western region

4.2.2 The U.S.A Soil Classification System

As mentioned in Chap. 2, after the study of Selkhozprom export, 1980, extensive soil survey studies have been conducted in different areas with medium to high potential for agriculture in Libya by different international companies or officers. In these studies, the U.S.A and FAO soil classification systems were used.

In the 1990s, Libyan soil scientists and researchers in the Department of Soil and Water Sciences at the University of

Table 4.2 The scheme of soil division into classes, types, subtypes, and genera in northeastern region

Soil type	Soil subtype	Soil genus
Red Ferrisiaallitic	Typical	Carbonate, carbonate saline, carbonate solonetzic-saline, leached, leached saline
	Concretionary	carbonate saline, leached, leached saline
	Crust	Carbonate, carbonate saline, leached, leached saline
	Hydrated	Carbonate, carbonate saline, leached, leached saline
	Hydromorphic of a truncated profile	carbonate solonetzic-saline, leached, leached saline
Yellow Ferrisiaallitic	Typical	leached, leached saline
	Concretionary	leached
Siallitic cinnamon	Typical	Carbonate, carbonate saline, leached
	Compact	Carbonate, leached
Rendzinas	Dark	Carbonate, carbonate saline, leached, leached saline
	Red	
Dark compact	Typical	Carbonate, carbonate saline
Reddish-brown arid	Differentiated	Carbonate, Carbonate saline, carbonate solonetzic-saline, leached, leached saline
	Differentiated crust	Carbonate, Carbonate saline, Carbonate Gypsic, Carbonate solonetzic, Carbonate solonetzic-saline
	Slightly Differentiated	Carbonate, Carbonate saline,
	Slightly Differentiated crust	Carbonate, Carbonate saline, Carbonate Gypsic
Reddish-brown arid	Non-Differentiated	Carbonate, Carbonate saline
	Hydromorphic crust	Carbonate saline, Carbonate solonetzic-saline
Brown arid	Differentiated	Carbonate, Carbonate saline, Carbonate solonetzic- saline
	Slightly Differentiated	Carbonate, Carbonate saline
Alluvial	Differentiated	Carbonate, Carbonate saline
Lithosols	Cinnamonic	Carbonate, Carbonate saline
	Reddish-brown	Carbonate, Carbonate saline, Carbonate solonetzic-saline
	Brown	Carbonate, Carbonate saline, Carbonate solonetzic-saline
Crusts	Monolithic	Carbonate, Siallitic Carbonate, Siallitic Carbonate saline
	Non-Monolithic	Carbonate, Carbonate saline,Siallitic Carbonate, Siallitic Carbonate saline
Solonchaks	Automorphic, Hydromorphic, Hydromorphic crust, Hydromorphic Sebeka	

Tripoli adopted the U.S.A soil classification system for all future soil survey work. At the same time, they decided to convert the available soil maps into the U.S.A Soil Taxonomy. Then, the Russian soil taxonomic units done in the northwest and northeast regions of Libya were converted to the U.S.A soil taxonomic units (Ben Mahmoud 1995). The main Libyan soil taxonomic units at soil order and soil suborder levels which resulted from soil studies are shown in Table 4.3 and Fig. 4.3 and Fig. 4.4

With the exception of Al Jabal Al Akhdarand Jabal Nafusa, all Libyan soils taxonomic units are either Entisols or Aridisols. Alfisols, Mollisols, and Vertisols are only found in Al Jabal Al Akhdar. Inceptisols are found in both Al Jabal Al Akhdarand Jabal Nafusa (Ben Mahmoud 1995).

The characteristics and distribution of Libyan soil taxonomic units on suborder and Great Group levels will be identified and discussed in Sect. 4.3.

4.3 Soil Taxonomic Units: Their Extent and Properties

4.3.1 Entisols

This soil order is considered as the most well-known type among Libyan soils, and it covers a large area in the northwestern, interior, and southern regions of Libya. Under this soil order, three suborders can be identified:

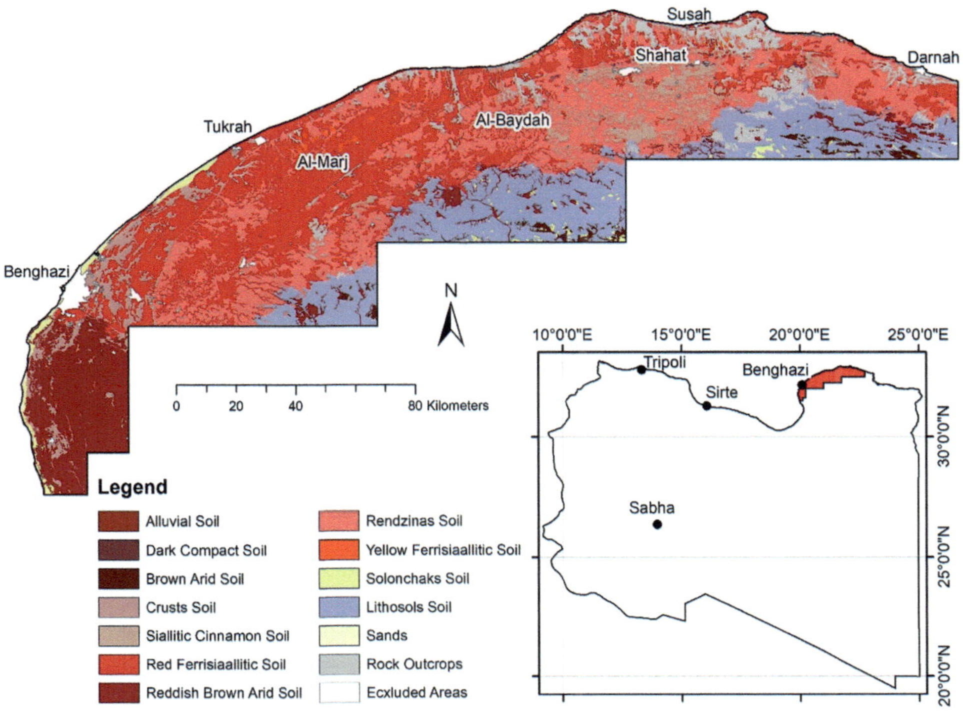

Fig. 4.2 Soil Types in the northeastern region

Table 4.3 The major U.S.A Soil Taxonomic Units in Libya (Ben Mahmoud 2013)

Soil Order	Soil suborder
Aridisols	Argids, Salids, Cambids, Calcids, Gypsids
Entisols	Psamments, Orthents, Fluvents
Inceptisols	Xerepts
Alfisols	Xeralfs
Mollisols	Rendolls
Vertisols	Xererts

- **(Psamments)**: These soils are defined as the soils in which the percentage of sand is more than 85% throughout soil profiles and the percentage of stones is less than 35%. In many cases, these soils consist of Aeolian sands and usually they are homogenous. They consist of one or two layers covered by an Ochric horizon on the surface and can be found in most of the Libyan lands. This type of soil is characterized by a newly formed soil profile that is not distinguished into subsurface horizons. The texture of this soil is sandy or loamy sand, and the depth of its profile ranges from deep to medium depth, although it may be shallow. The structure of these soils is mainly structureless (single grains). The degree of reaction tends to be alkaline (pH more than 7), and the amount of calcium carbonate ranges from low to medium. In addition, their organic matter content is low.

These soils suffer from a decrease in their ability to hold water; as it has a high infiltration rate, it also suffers from a decrease in its nutritional content, especially nitrogen, phosphorus, and trace elements, and are highly vulnerable to soil erosion by wind. The total estimated area covered with these soils is about 2130000 hectares. These soils are classified under the great group level in Libya according to the dominant Moisture Regime (Table 4.4 and Fig. 4.5) (Ben Mahmoud 2013).

(This picture was taken from Aziziya area, south of Tripoli, by Prof. Ben Mahmoud, it is a sandy textured, very deep, reddish-brown, it has only ochric horizon, land use: field crops, vegetables, and fruit trees).

- **(Orthents)**: Generally, these soils are shallow or moderately deep and contain the "Ochric" horizon at the surface followed directly by the rocky bed or by hard rocky layer. The stone percentage can occasionally reach more than 35%. Some of these soils are gravelly textured (containing more than 20% of their dry weight) and including solid components with diameters greater than

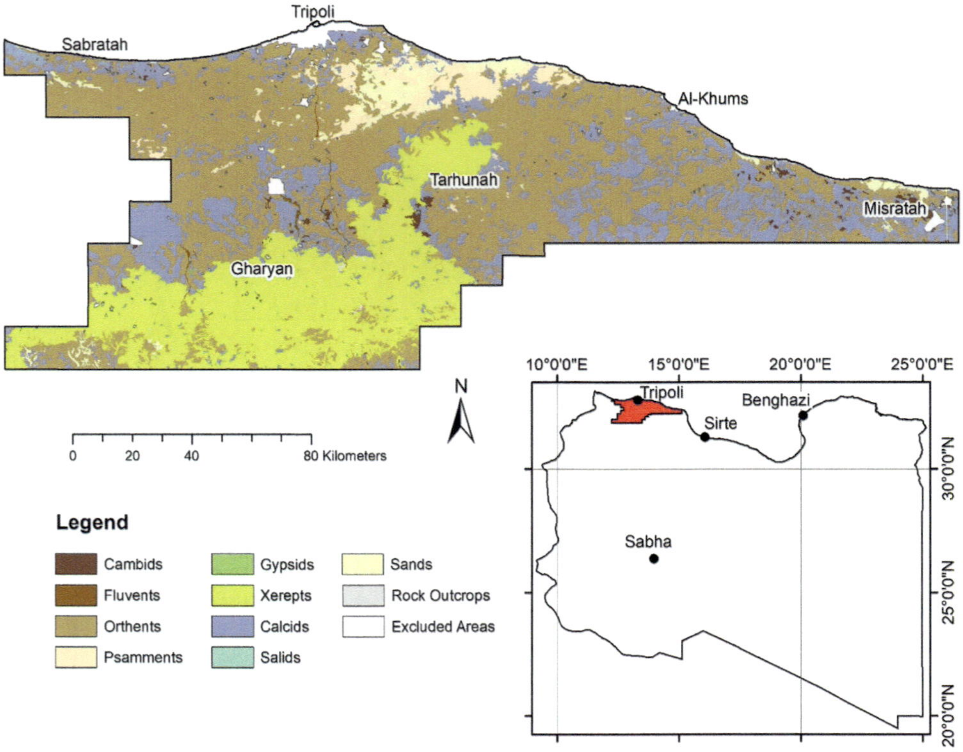

Fig. 4.3 The Map of Soil Suborder in northwestern region

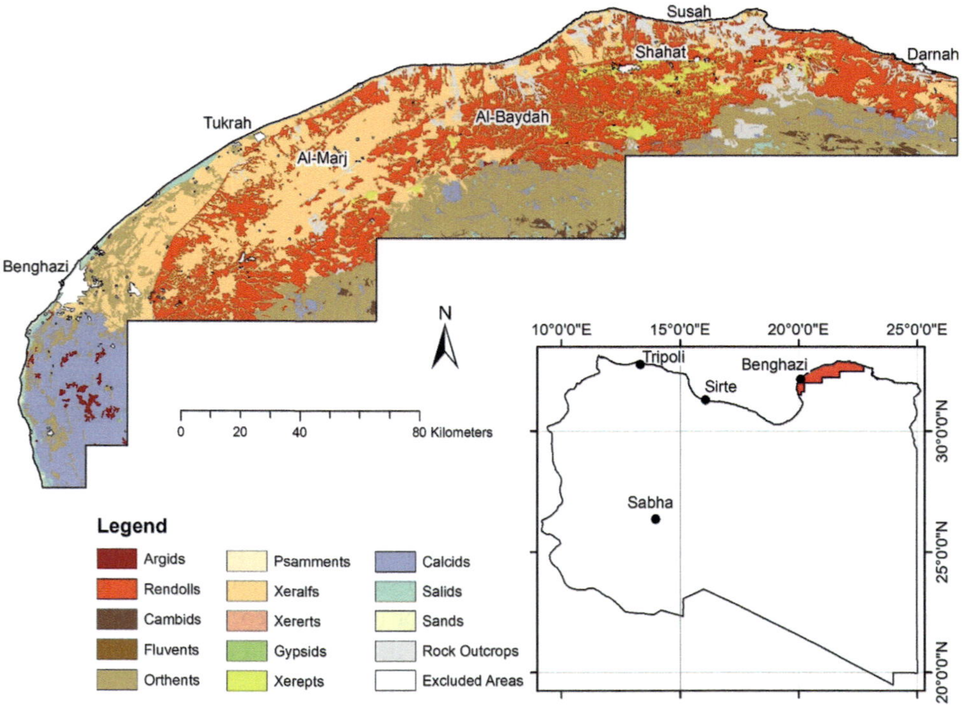

Fig. 4.4 The Map of Soil Suborder in northeastern region

Table 4.4 Classification of Psamments (Ben Mahmoud 2013)

Order	Suborder	Great group
Entisols	Psamments	Xeropsamments
		Torripsamments

Fig. 4.5 Torripsamments (This picture was taken from Aziziya area, south of Tripoli by Prof. Ben Mahmoud, it is a sandy textured, very deep, reddish brown, it has only ochric horizon, land use: field crops, vegetables, and fruit trees)

Table 4.5 Classification of Orthents (Ben Mahmoud 2013)

Order	Suborder	Great groups
Entisols	Orthents	Xerorthents
		Torriorthents

2 mm and less than 75 mm) on their surface or within their soil profile. They can also be stony soils (in which the diameters of these solid components exceed 75 mm) containing rocks on the surface or buried in different depths of the soil profile. The presence of these solid components in soils of whatever size leads to the obstruction of various service operations of plowing, hoeing, and others, as well as hindering seedling growth and stopping or limiting the movement of plant roots. These soils are present in many regions, especially in mountainous areas, southern and desert regions; the total estimated area covered with these soils is about 532500 hectares. These soils are classified under the level of a great group in Libya according to the dominant Moisture Regime (Table 4.5 and Fig. 4.6) (Ben Mahmoud 2013).

Fig. 4.6 Torriorthents (This picture was taken from Al Assah area, south-west of Tripoli by Prof. Ben Mahmoud, it is a gravelly loamy sand textured, moderately deep, brown, it has only ochric horizon, land use: mainly range lands and in limited areas cultivated by rainfed barley)

(This picture was taken from Al Assah area, southwest of Tripoli, by Prof. Ben Mahmoud, it is a gravelly loamy sand textured, moderately deep, brown, it has only ochric horizon, land use: mainly rangelands and in limited areas cultivated by rainfed barley).

- **(Fluvents)**:

These are the soils of alluvial old dry valleys (Wadis) and their outlets. During the formation process, the soil particles are transported and deposited in water. Generally, they consist of alluvial layers, and each of these layers represents a stage of valley flood. Initially, the original materials are formatted (deposited), followed by the formation of the Ochric horizon on the surface, which can only occur after a period of time succeeding its stabilization. The total estimated area of these soils is about 106500 hectares. The characteristics of these soils differ according to the different geological materials carried by water, the intensity of the surface runoff, and the topography of the valley. Fluvents are mostly brownish to reddish soils that are formed in old or recent water deposited sediments. The ages of these layers differ from one soil to another, and can even differ within the same soil, and are often interspersed with layers deposited by wind. Sometimes these soils are buried under a shallow layer of clear sand deposited by active wind erosion. The most important characteristics that unite most of the soils of different sedimentary valleys are their dark color, which is often a dark brown, as well as their heavy, often clayey-loamy texture and varying proportions of gravel. The degree of reaction in these soils tends to be alkaline. The percentage of salinity in it varies from one soil to another, and is often from simple to medium, as it contains varying amounts of calcium carbonate and gypsum. The fertility status of these soils is, like other soils in dry areas, poor in organic matter and some nutrients, especially nitrogen and phosphorus. These soils are classified under the great group level in Libya, according to the dominant Moisture Regime (Table 4.6 and Fig. 4.7) (Ben Mahmoud 2013).

(This picture was taken from wadi SofelJeen area, south of Sirte, by Prof. Ben Mahmoud, it is a loamy or sandy loam textured, deep soil, dark brown, it has only ochric horizon, land use: rainfed barely).

Table 4.6 Classification of Fluvents (Ben Mahmoud 2013)

Order	Suborder	Great groups
Entisols	Fluvents	Xerofluvents
		Torrifluvents

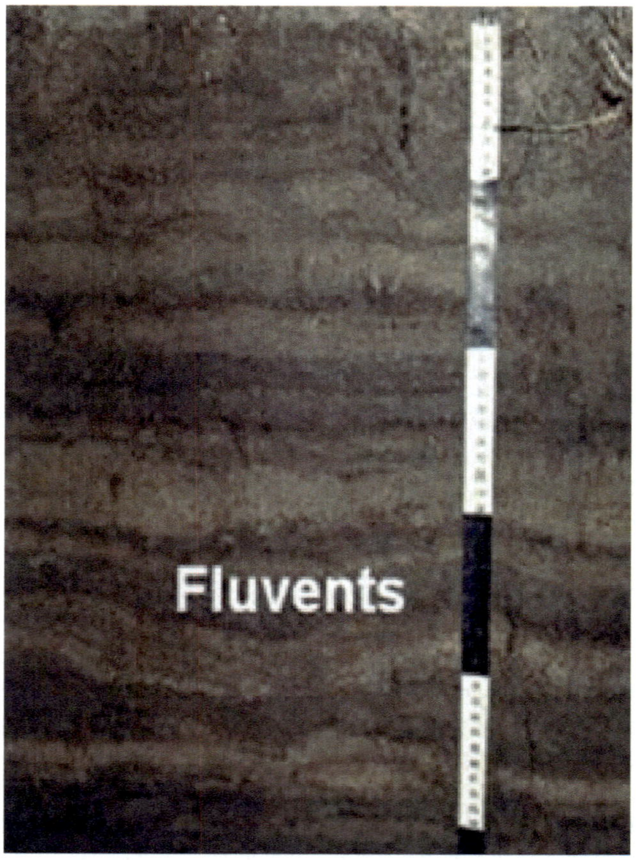

Fig. 4.7 Torrifluvents (This picture was taken from wadi SofelJeen area, south of Sirte by Prof. Ben Mahmoud, it is a loamy or sandy loam textured, deep soil, dark brown, it has only ochric horizon, land use: rainfed barely)

4.3.2 Aridisols

These soils are the second most common soils in Libya and can be found in the desert and semidesert regions (Ben Mahmoud 2013). Below are the main suborders of the Aridisols in Libya:

- **(Calcids)**: These soils have a pronounced accumulation of calcium carbonate ($CaCO_3$), either as a continuous layer or as gathered particles. These soils are characterized by their undeveloped soil profile that contains the diagnostic subsurface horizon called the Calcic or Petrocalcic horizons in which the calcium carbonate transferred to it from the horizons above it has accumulated along with the soil profile. Calcids are usually nonsaline and can be identified by soil profiles of varying depths and low content of organic matter and nitrogen. It is also poor in phosphorus and trace elements available for plant absorption. Calcids can also be found in the western and central coastal areas, in the northwestern highlands, and is also found in some northern and southern old valleys, and the southern desert regions. The total estimated area covered with these soils is 594000 hectares. These types of soils exist in all county regions without any exception, and they are classified under the level of the great group as shown in Table 4.7 and Fig. 4.8a, b (Ben Mahmoud 2013).

(a). This picture was taken from Wadi Al Athal area, south of Zawiya, by Elhawej, it is a loamy or clay loam textured, deep soil, light brown, high in $CaCO_3$ throughout the profile, it has Ochric and Calcic horizons, land use: rainfed wheat and barley).

(b). This picture was taken from Dirj area, near Gadamis, by Prof. Elhawej, it is a sandy loam textured, moderately

Table 4.7 Classification of Calcids (Ben Mahmoud 2013)

Order	Suborder	Great group
Aridsols	Calcids	Haplocalcids
		Petrocalcids

(a)

(b)

Fig. 4.8 **a** Haplocalcids (This picture was taken from Wadi Al Athal area, south of Zawiya by Elhawej, it is a loamy or clay laom textured, deep soil, light brown, high in $CaCO_3$ throughout the profile, it has Ochric and Calcic horizons, land use: rainfed wheat and barley) and **b** Petrocalcids (This picture was taken from Dirj area, near Gadamis by Prof. Elhawej, it is a sandy loam textured,moderately deep soil, light brown, high in $CaCO_3$ throughout the profile. It has Ochric and Petrocalcic horizons, present land use: rangeland)

Table 4.8 Classification of Gypsids (Ben Mahmoud 2013)

Order	Suborder	Great group
Aridsols	Gypsids	Haplogypsids
		Calcigypsids
		Petrogypsids

deep soil, light brown, high in $CaCO_3$ throughout the profile. it has Ochric and Petrocalcic horizons, present land use: rangeland)

- **(Gypsids)**: Gypsids are commonly found in Libya, in an arid, and semiarid climate. They get their name from the accumulation of gypsum salt that forms what is known as the gypsic horizon. Gypsum may be present in this soil in the form of formations of various shapes and sizes, in the form of scars, tubes, or gypsum veins. The gypsum horizon may solidify in some types of these soils and form a gypsum deaf layer known as the petrogypsic horizon. These soils are usually saline as well. The total estimated area covered with these soils in Libya is about 124250 hectares. They are commonly spread in some desert depressions such as the oasis of Ghadames, Daraj, Jufra, Tawergha, and Ajdabiya, and they are classified under the great group level of Libyan soils as shown in Table 4.8 and Fig. 4.9a, b (Ben Mahmoud 2013).

(a) This picture was taken from Jafara area, southern part of Libya, by Prof. Elhawej, it is a loamy or clay loam textured, deep soil, light brown, high in gypsum throughout the soil profile, it has Ochric and gypsic horizons, land use: palm trees).

(b) This picture was taken from Tawergha area, north-central part of Libya, by Prof. Ben Mahmoud, it is a clay loam textured, moderately deep soil, light brown, high in gypsum throughout the soil profile, it has Ochric and Petrogypsic horizons, present land use: rangeland).

- **(Salids)**: Salids are the salty soils that can be found in many regions near the Libyan coast or in some southern desert regions. The salts in these soils are accumulated as a salty horizon known as (Salic horizon) inside the soil profile. This type of soil is characterized by a large amount of dissolved salts in the water, which impede or prevent the normal growth of agricultural crops. Salids (sometimes referred to as Saline or Sebeka soils by locals) are divided into two main types; the first is called Hydromorphic, which spreads in semiarid areas and consists of materials rich in salts. This type of Salids can be found mainly in the northeastern region of Libya, where the groundwater level is relatively deep.

The second and more widespread type of Salids in Libya is called Automorphic, in which the low local topography and the high water table level play the major roles in its formation. In the flat depressions regions, generally close to the water table with high temperatures, the evaporation of water occurs rapidly from the soil surface, pushing the movement of capillary water from the top of the groundwater level to the surface of the soil. Here, the water evaporates leaving the salts flowering on the surface layer of the soil, and as this process continues, a gradual accumulation of salts occurs, forming a distinctive salt cover known as the "salt crust" (see Fig. 1.7, Chap. 1).

Fig. 4.9 **a** Haplogypsids (This picture was taken from Jafara area, southern part of Libya by Prof. Elhawej, it is a loamy or clay loam textured, deep soil, light brown, high in gypsum throughout soil profile, it has Ochric and gypsic horizons, land use: palm trees) **b** Petrogypsids (This picture was taken from Tawergha area, north-central part of Libya by Prof. Ben Mahmoud, it is a clay loam textured, moderately deep soil, light brown, high in gypsum throughout soil profile. It has Ochric and Petrogypsic horizons, present land use: rangeland)

Table 4.9 Classification of Salids (Ben Mahmoud 2013)

Order	Suborder	Great group
Aridsols	Salids	Haplosalids
		Aquisalids

One of the important things that should be known about the Saline and Sebeka soils is the depth of the salt horizon. The closer it is to the surface, the more harmful it is to the growth of plants. Salty soils usually contain several types of salts prevailing and they vary in the extent of damage they cause to plants. The most harmful salt to plants is sodium salt, followed by chloride salts, while the least harmful is sulfate salts. Where the concentration of sodium salts is high relative to other types of salt, a sodic soil may develop (Zurqani el at. 2018, 2019).

These soils also differ in terms of textures: some of them are sandy soils, some silt sand, and others clay loam, but all are characterized by their weak integration and severity of disintegration. They are spread in the western and central coastal areas, the Benghazi plain region, and most of the oases and southern old valleys, with a total estimated area of about 400,000 hectares. These soils are classified under the great group level in Libya based on the existence or absence of a groundwater level that is close to the surface (Table 4.9 and Fig. 4.10) (Ben Mahmoud 2013).

(This picture was taken from south-west of Benghazi, by Prof. Ben Mahmoud, it is a clay loamy textured, moderately deep soil, Reddish-brown, high in salts (more than 2%) throughout the profile, presence of water table, it has Salic horizon, present land use: rangeland for camels)

Fig. 4.10 Haplosalids (This picture was taken from south-west of Benghazi, by Prof. Ben Mahmoud, it is a clay loamy textured,moderately deep soil, Reddish brown, high in salts (more than 2%) throughout the profile, presence of water table. it has Salic horizon, present land use: rangeland for camels)

Table 4.10 Classification of Cambids (Elhawej and Elaalem 2012)

Order	Suborder	Great group
Aridsols	Cambids	Haplocambids

- **(Cambids)**: Cambids are soils that contain an alterable horizon (Cambic horizon) under the surface, and an Ochric surface horizon usually exists in these soils as well. Cambids are present in some regions of the Jafara Plain such as Al Aziziye, Kaser Bin Ghashir, and some areas in the Nafusa mountains range, where the geomorphologic surfaces are relatively stable. These soils are characterized by the texture of sandy loam or loamy sand and are mostly deep to medium depth. They are also almost completely free from excessive salinity and contain a small percentage of calcium carbonate throughout the soil profile. Cambids are more suitable than psamments and orthents, but they are also poor in organic matter and nutrients. There is only one great group level of these soils in Libya (Table 4.10 and Fig. 4.11) (Ben Mahmoud 1995; Elhawej and Elaalem 2012).

Fig. 4.11 Haplocambids (This picture was taken from Qasr bin Ghashir in the Jafara Plain by Prof. Ben Mahmoud, it is a sandy loam or loamy sand textured, very deep soil, Reddish brown, non-saline with low content of $CaCO_3$. It has Ochric and cambic horizon, present land use: cultivated by fruit trees and vegetables)

(This picture was taken from Qasr bin Ghashir in the Jafara Plain by Prof. Ben Mahmoud, it is a sandy loam or loamy sand textured, very deep soil, Reddish-brown, non-saline with low content of $CaCO_3$. It has Ochric and cambic horizon, present land use: cultivated by fruit trees and vegetables).

- **(Argids)**: Argids are soils in which a clay horizon is formed due to one of the two reasons: The first reason is that the soil is old, it was formed in old clay deposits and stable geomorphology, and under an ancient climate in the region is much wetter than it is now. Another reason the horizon may be formed is the rainfall during winter and autumn where a high average amount of rainfall resulted in weathering, releasing aluminum silicate minerals, and producing the clay. The clay was transported under the surface making the clay horizon (Argillic horizon). These soils are characterized by the clay texture and a soil profile that is mostly moderately deep to deep. As for their other properties, they are differentiated by their mostly low levels of salinity and gypsum, but sometimes high levels of Exchangeable sodium percentage (ESP). The ability of these soils to hold water is high, but they are lacking in organic matter like nitrogen, phosphorus, and trace elements while being rich in basic nutrients such as potassium, calcium, and magnesium. In Libya, these soils are classified under the great group level as shown in Table 4.11 and Fig. 4.12 (Ben Mahmoud 2013).

(This picture was taken from Benghazi Plain, by Prof. Ben Mahmoud, it is a clayey textured, very deep to deep soil, Reddish-brown, nonsaline with low content of $CaCO_3$. It has Ochric and argillic horizon, present land use: cultivated by wheat).

Table 4.11 Classification of Argids (Ben Mahmoud 2013)

Order	Suborder	Great group
Aridsols	Argids	Paleargids
		Haplargids
		Natrargids
		Durargids

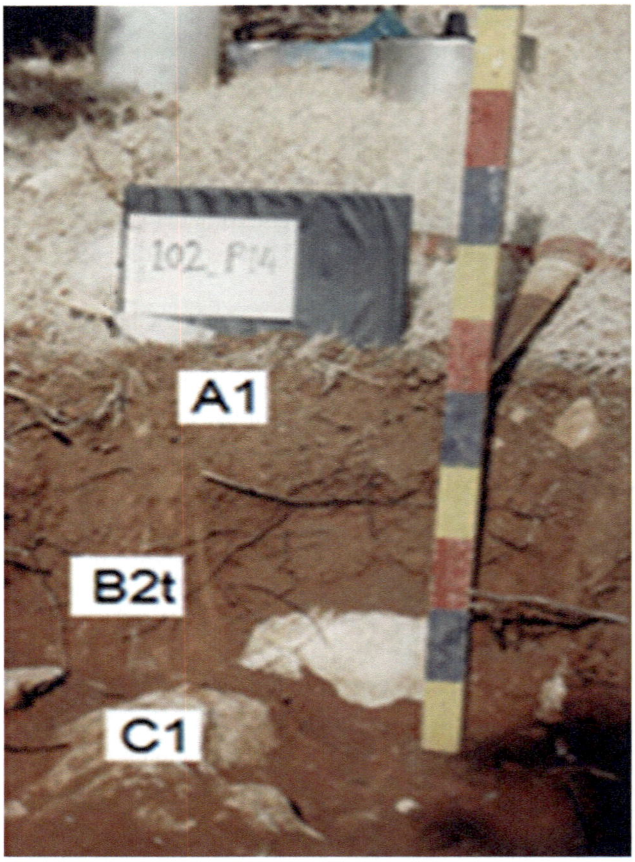

Fig. 4.12 Haplargids (This picture was taken from Benghazi Plain by Prof. Ben Mahmoud, it is a clayey textured, very deep to deep soil, Reddish brown, non-saline with low content of $CaCO_3$. It has Ochric and argillic horizon, present land use:cultivated by wheat)

Fig. 4.13 Rhodoxeralfs (This picture was taken from Al Marj in the Al Jabal Al Akhdar by prof. Ben Mahmoud, it is a clayey textured,very deep soil, Red in color, non-saline with low to moderate content of $CaCO_3$. It has the Ochric and Argillic horizon, present land use: cultivated by wheat)

4.3.3 Alfisols

This soil order can be found only in the northeast region of Libya such as: Al Marj, and in the Plains of Al Jabal Al Akhdar"Green Mountain". Below is the main classified suborder of these soils in Libya:

- **(Xeralfs)**: These soils are known in the Mediterranean region as "Terrarossa" and are distinguished by their reddish color patterns. They are also characterized by an Ochric surface horizon and a rate of base saturation at least 35% of the total saturation rate. In Libya, these soils are present in the region of Al Jabal Al Akhdar "Green Mountain" in the areas that receive more than 350 mm of precipitation per year. These areas are considered within the semiarid to semihumid regions and are covered with forest vegetation known as Mediterranean flora. The soil profile of these soils is relatively developed compared to other soils in Libya. It is also mostly deep, and in some areas, it may reach as deep as 3 meters. However, in other areas, we find it is characterized by the presence of a shallow soil profile. These shallow soils are often found in sloping areas or those that are subject to severe soil erosion by water. Xeralfs are also characterized by the presence of the Argillic horizon with a polygonal mass structure. In addition to their red color, these soils are distinguished by their clay texture, where the percentage of clay particles reaches more than 35%, and thus their ability to hold water is high.

Table 4.12 Classification of Xeralfs (Ben Mahmoud 2013)

Order	Suborder	Great group
Alfisols	Xeralfs	Rhodoxeralfs
		Natrixeralfs
		Durixeralfs
		Haploxeralfs

As for their other properties, these soils are mostly low in salinity and gypsum, but they differ in the extent to which they contain calcium carbonate. Some of which are completely free of calcium carbonate, while others contain high levels of it, sometimes reaching 20% or more. They are poor in organic matter and nutrients such as nitrogen, phosphorus, and trace elements, but rich in basic nutrients such as potassium, calcium, and magnesium. In Libya, these soils are covered by approximately 356,000 hectares and are classified under the great group level as shown in Table 4.12 and Fig. 4.13 (Ben Mahmoud 2013).

(This picture was taken from Al Marj in the Al Jabal Al Akhdar, by prof. Ben Mahmoud, it is a clayey textured, very deep soil, Red in color, nonsaline with low to moderate content of $CaCO_3$. It has the Ochric and Argillic horizon, present land use: cultivated by wheat).

4.3.4 Mollisols

These soils can be found only in the Al Jabal Al Akhdar and they are classified under a single suborder level in Libya (Ben Mahmoud 1995).

- **(Rendolls)**: These soils are known in the old classification systems as "Rendzina". They are present in flat low regions of the Al Jabal Al Akhdar, where the water gathers on the surface. Accordingly, a dense vegetation cover of grass and natural plants is growing on these soils for long periods during the year. These conditions result in a noticeable accumulation of the organic materials which consequently cause the formation of the organic mineral horizon (Mollic Epipedon). In many instances, the profile of these soils contains limestones that are close to the surface or hard calcic layers since these soils originally formed from limestone. These soils are characterized by a shallow soil profile, as their depth does not exceed 60 cm. There are two types of these soils: the first is reddish and the second is dark gray. It was observed that the color of the formed soil depends on the original material, as pure hard limestone forms red soil, while soft limestone containing impurities gives dark gray soils (Ben Mahmoud 1995).

These soils are characterized by a high percentage of calcium carbonate and a relatively high percentage of organic matter compared to the rest of Libyan soils, being comprised of about 2% organic materials. The texture of these soils ranges from loamy to clay loam, and the degree of their reaction tends to be alkaline (Fig. 4.14 and Table 4.14).

These soils occupy vast areas in the Al Jabal Al Akhdar region, covering an area of approximately 340,000 hectares. It is located mainly in Sultanah, Sousse, Al Marj, Al Bayda, Qasr Libya, and other regions. In Libya, only a single great group of these soils was identified as shown in (Table 4.13 and Fig. 4.13) (Ben Mahmoud 2013).

Fig. 4.14 Haprendolls (This picture was taken from Qasr Libya in the Al Jabal Al Akhdar by Prof. Ben Mahmoud, it is a loam or clay loam textured, shallow soil, the bedrock is limestone, dark in color, relatively high in organic matter, nonsaline with a high content of $CaCO_3$. It has a Mollic horizon, present land use: The use of these soils in agriculture is very limited and that is why they are left as they are in nature (natural forest trees and bushes)

Table 4.13 Classification of Rendolls (Ben Mahmoud 2013)

Order	Suborder	Great group
Mollisols	Rendolls	Haprendolls

4.3.5 Inceptisols

These soils cover limited small areas in Libya and can be found in the mountain areas in both (Al Jabal Al Akhdar, as well as Gharyan, Tarhuna and Yafran of the northwestern mountains), and only one suborder of these soils was observed:

- **(Xerepts)**: These soils cover limited small areas in Libya which are estimated of about 257 thousand hectares. These soils are characterized by a slight development in the soil profile, as they contain a cambic horizon (B horizon). They also contain very clear modern formations of calcium carbonate concentrations in the form of spots, scars, and veins spread throughout the soil profile. Among its most important characteristics is its color, which is cinnamonic. As for their other properties, these soils are deep to moderately deep, the texture ranges from sandy loam to loam, the soils are mostly low in salinity and gypsum, But, they differ in the extent to which they contain calcium carbonate (medium to high). These soils are poor in organic matter and nutrients such as nitrogen, phosphorus, and trace elements, but rich in basic nutrients such as potassium, calcium, and magnesium (Table 4.14 and Fig. 4.15) (Ben Mahmoud 2013; Sherif 1995).

(This picture was taken from Tarhuna, south of Tripoli, by Prof. Ben Mahmoud, it is a sandy loam or loam sand textured, deep to very deep soil, cinnamonic in color, nonsaline with moderate content of $CaCO_3$, it has Ochric and Cambic horizons, present land use: usually grown olives, figs, and almonds).

Fig. 4.15 Calcixerepts (This picture was taken from Tarhuna, south of Tripoli by Prof. Ben Mahmoud, it is a sandy loam or loam sand textured, deep to very deep soil, cinnamonic in color, non-saline with moderate content of $CaCO^3$. It has Ochric and Cambic horizons, present land use, usually grown olives, figs and almonds)

4.3.6 Vertisols

These soils are distinguishable by the presence of cracks due to the shrink and swell of clay minerals such as the "Montmorillonite", and only one suborder of these soils was observed:

- **Xererts**: This type of soil is limited to small areas of Al Jabal Al Akhdar, estimated at 1,544 hectares. It is found in the Al Bayda and Al Qobba area, near Al Fadia. In areas where this type of soil is spread, rainfall rates reach more than 450 mm per year. The predominant color of these soils is dark gray, and the texture is clayey, in which the clay mineral montmorillonite predominates. This material is what expands and contracts as a result of wetness and drought to give distinct cracks.

One of the most important characteristics of these soils is that the horizons are not distinguished in soil profiles, and they often contain cracks during periods of drought, with a width sometimes reaching more than 3 cm and a depth of more than 1 m. The cracking phenomenon disappears in the winter during the wet period. These cracks play a distinct role in creating the characteristic wavy topography on the surface, as soil particles move from the surface to the inside of the soil through cracks, ultimately leading to poor recognition of the horizons within their profiles. As a result, this soil was called the "stirred soil", a term for Vertisols still used around the world today.

This soil contains small proportions of organic matter that does not exceed 1%. Its degree of reaction tends to be alkaline, and the percentage of calcium carbonate in it is high. Its natural properties are considered poor, as it is highly clayey in texture with low permeability and poor internal drainage, but the exchange capacity of cations is high. As for its degree of fertility, it is poor in nitrogen, phosphorus, and trace elements, but rich in other nutrients such as potassium, calcium, and magnesium. Among the

Table 4.14 Classification of Xerepts (Ben Mahmoud 2013)

Order	Suborder	Great group
Inceptisols	Xerepts	Calcixerepts
		Haploxerepts
		Durixerepts

Table 4.15 Classification of Xererts (Ben Mahmoud 2013)

Order	Suborder	Great group
Vertisols	Xererts	Haploxererts
		Durixererts
		Calcixererts

Fig. 4.16 Haploxererts. (This picture was taken near Al Bayda city in Al Jabal Al Akhdar by Prof. Ben Mahmoud, it is a clayey textured,deep to moderately deep soil, dark in color, non-saline with a high content of $CaCO^3$. The soil has deep and wide cracks, with a wavy surface topography. Its horizons are not distinguished in the profile, present land use: usually avoid cultivate it by trees but can be grown rainfed wheat)

problems of cultivating these soils are cracking and difficulty in serving (Ben Mahmoud 2013). Table 4.15 shows its Great Groups found in Libya and Fig. 4.16 shows a picture of its soil profile.

(This picture was taken near Al Bayda city in Al Jabal Al Akhdar, by Prof. Ben Mahmoud, it is a clayey textured, deep to moderately deep soil, dark in color, nonsaline with a high content of $CaCO_3$. The soil has deep and wide cracks, with a wavy surface topography. Its horizons are not distinguished in the profile, present land use: usually avoid cultivate it by trees but can be grown rainfed wheat).

4.4 Conclusions

Libya is characterized by a desert environment, where the majority of the Libyan lands are covered with non-soil. These areas include Sands and Sand Dunes, Rocky Deserts (Hammadas), Desert Pavements (Sarrier), and Saline Depressions (Sebkha). Entisols and Aridisols are by far the most extensive soil orders in the country, followed by lesser areas of Alfisols, Mollisols, and Vertisols which can only be found in Al Jabal Al Akhdar. Inceptisols are placed in small areas both Al Jabal Al Akhdarand Jabal Nafusa. Generally, the majority of Libyan soils are undeveloped or partially developed with low levels of organic matter, which is typical of the arid and semiarid regions. Furthermore, these soils are suffering from problems and obstacles that reduce their agricultural capability (i.e., erosions, salinity, unsuitable soil texture, shallow soil profile, and low soil fertility). Therefore, sustainable use and appropriate soil conservation practices are needed, which will lead to more efficient use of soil resources and achieve future food security in the country.

References

Ben Mahmoud KR (1995) Libyan soils (Their Genesis, Classification, Properties and Agricultural potentials) NASR, Tripoli, Libya, pp 615 (in Arabic)

Ben Mahmoud KR (2013) Towards a national strategy for the sustainability of natural resources and enhancing food security in Libya, pp 123 (in Arabic)

Elhawej AZ, Elaalem MM (2012) A guide morphological description and classification soil profile in the field, pp 80 (in Arabic)

Selkhozpromexport (1980) Soil studies in the Eastern zone of thesocialist people's Libyan Arab Jamahiriya. In: Secretariat for Agricultural Reclamation and Land Development,Tripoli

Sherif AS (1995) The nature, composition and classification of mountain soil in Gharyan, Yafran and Tarhuna. Published Master thesis, University of Tripoli. (in Arabic)

Solkhozprom Export. Soil Ecological Expedition. USSR 1980. Soil studies in western zone, eastern zone and central zone of Libya. Ministry of Agr. and Land Development. Tripoli. Three separate volumes. (in English)

Zurqani HA, Mikhailova EA, Post CJ, Schlautman MA, Elhawej AR (2019) A review of libyan soil databases for use within an ecosystem services framework. Land 8(5):82

Zurqani H, Mikhailova E, Post C, Schlautman M, Sharp J (2018) Predicting the classes and distribution of salt-affected soils in north-west Libya. Commun Soil Sci Plant Anal 49(6):689–700

Dr. Mukhtar M. Elaalem is an associate professor of soil pedology and head of the Department of Soil and Water Sciences, Faculty of Agriculture, University of Tripoli, Tripoli, Libya. He got a Ph.D. degree in Geographic Information System at Leicester University, UK in 2011. His Ph.D. thesis is on "The Application of Land Evaluation Techniques in Jafara Plain in Libya using Fuzzy Methods", and his research interest is soil pedology, geographical information science, natural resources applications of GIS (especially land evaluation and soils studies), modeling spatial uncertainty, especially employing fuzzy logic methods to land evaluation and soils studies. He teaches various courses, including introduction to soil science, soil survey; genesis and classification, and Libyan soils. He attended many local, regional, and international meetings, workshops, training programs, conferences on land resources, and applications of GIS and remote sensing.

Dr. Hamdi A. Zurqani is one of the faculty members at the Department of Soil and Water Sciences, Faculty of Agriculture, University of Tripoli, Tripoli, Libya. Dr. Zurqani is a recognized expert as a result of his internationally acclaimed work in the areas of environmental information science, remote sensing, land evaluation, sustainability, pedology, and soil science education. He has conducted research across the world, including the United States of America and Africa. Dr. Zurqani is a distinguished soil scientist with a wide range of scientific and working experiences in Libya and abroad. He received his M.Sc. (2010) from the University of Tripoli, Tripoli, Libya and Ph.D. (2019) from Clemson University, Clemson, SC, USA. His major research and teaching activities at the University of Tripoli have focused mainly on soil genesis and classification and the environmental information sciences (Remote Sensing and Geographic Information System). He has published broadly in many journals (e.g., Nature "Scientific Reports", Geoderma; International Journal of Applied Earth Observation and Geoinformation; Journal of Geological Sciences; Land; Frontiers in Environmental Science; Communications in Soil Science and Plant Analysis; and others). Dr. Zurqani is a member of the Editorial Board for Remote Sensing (MDPI) Journal, counseling outcome and research evaluation. He also was appointed to serve as a Guest Editor for the Special Issue "Applications of Remote Sensing in Earth Observation and Geo-Information Science". In addition, Dr. Zurqani conducted peer-review for many journals including Journal of Environmental Informatics, Applied Sciences, SN Applied Sciences, Remote Sensing, Heliyon, Geosciences, Land, Water, Agronomy, Agriculture, Sustainability, Arid Land Research and Management, International Journal of Environmental Research and Public Health, Natural Hazards, and Conference of the Arabian Journal of Geosciences. He is also one of the authors of the lab manual entitled, "GIS Exercises for Natural Resource Management". Dr. Zurqani has been the recipient of numerous awards and honors: Recipient of Douglas R. Phillips Award for Graduate Students, Department of Forestry and Environmental Conservation, Clemson University, April 12, 2019; the First Place Best Judged Poster (CAFLS) at the GRADS 2019: Clemson Student Research Forum on April 4, 2019; the Second Place Poster at the 11th Clemson Biological Sciences Annual Student Symposium, April 6, 2019; the Second Place Best Judged Poster at the Clemson Student Research Forum on April 4, 2018; and the Third Place Poster at the 9th Clemson Biological Sciences Annual Student Symposium, February 25, 2017. Dr. Zurqani conducts cutting-edge research in the field of environmental information science, remote sensing, land use management/planning, change detection of landscape degradation, and geographic information system (GIS) models. He has focused on his research efforts on the development of new technologies in the field of environmental information sciences, geo-intelligence (advanced geo-information science and earth observation, machine and deep learning, and big data analytics), remote sensing, land evaluation, pedology, land use management/planning, monitoring and evaluating sustainable land management, change detection of landscape degradation, and geographic information system models.

Dr. Khaled R. Ben Mahmoud is a senior professor of soil pedology. He is a distinguished soil scientist with wide range of scientific and working experiences in Libya and abroad. He received his Ph.D. in 1977 from Colorado State University (USA). His major teaching activities and research are at the Department of Soil and Water, Faculty of Agriculture, University of Tripoli, Libya. He has been teaching undergraduate and graduate students in Libya and has supervised many M.Sc. and Ph.D. thesis and dissertation in Libya and abroad. He focused mainly on soil resources, land evaluation, and mapping. He contributes to soil science in Libya and abroad as the Head of the Department of Soil and Water Sciences, Faculty of Agriculture, University of Tripoli, Tripoli, Libya (1978–1983), Dean of the Faculty of Agriculture, University of Tripoli, Tripoli, Libya (1983–1986), Chairman of the Committee for Libyan Agricultural Mapping, Tripoli, Libya (1988–1990). A Research Coordinator, University of Tripoli, Tripoli, Libya (1997–2001), Member of the Scientific Council of Agricultural Research Center (ARC), Tripoli, Libya (2000–2002), Member of the Scientific Council of the College of Higher Studies at the University of Tripoli, Tripoli, Libya (2000–2002), Member of the Scientific Committee of Libya Centre for Remote Sensing and Space Sciences, Tripoli, Libya (1998–2006), A National Coordinator of the Project implemented by FAO: Mapping of Natural Resources for Agricultural Use and Planning Tripoli, Libya (2000–2006), A General Coordinator of the National Programme for the Development of Vegetation Cover, Tripoli, Libya (2004–2006). Director of the Division of Land and Water Uses, The Arab Centre of the Studies of Arid Zones Lands (ACSAD), The League of Arab States, Damascus, Syria (2007–2011), Head of the Programme: Monitoring and Combating of Desertification in the Arab World, The Arab Centre of the Studies of Arid Zones Lands (ACSAD), The League of Arab States, Damascus, Syria (2007–2011), Director General of Eddayra Co. (Agricultural and Environmental Consulting and Services), Tripoli, Libya (2011–to date). Engaged as a local and international expert to control, evaluate, approve, and accept of many soils studies and mapping which have been implemented by different foreign and local companies, agencies, and offices in different parts of Libya and abroad (1977–to date). More than 100 country reports, research and scientific papers, and other publications were published in different local and international scientific journals and symposiums (in Arabic and English). He is the author of more than 10 books that have been written in Arabic. The most important ones are: Sandy soils of Libya (1981), The study of soils in the field (1984), Principles of Soil Science for Agricultural Secondary and Technical Schools (1985), Fundamentals of Soil Science and its Relation to Plant Growth (1995), Libyan Soils (Their Genesis, Classification, Properties, and Agricultural Potentiality (1995), Fundamentals of Soil and water Science (2001), Issues on Desertification and Land Degradation in the Arab Region (2010), Towards a National Strategy for the Sustainability of Natural Resources and Enhancing Food Security in Libya (2013), and The Libyan Desert (its natural potentials and prospects for investment) in press. Finally, he was appointed as a technical advisor to the President of the Libyan Presidential Council for Agriculture, Tripoli, Libya (2018–to date).

Dr. Azzeddin R. Elhawej is a professional soil scientist with a wide range of scientific and practical experiences in Libya and Africa. He holds master's degree in science (1974) from Arizona State University, Tucson, Arizona, USA, and a Ph.D. (1980) from Oklahoma State University, Stillwater, Oklahoma, USA. His main research and teaching activities at University of Tripoli focused on soil genesis and classification, clay mineralogy, and soil mapping. He has contributed to the advancement of soil science in Libya as Head of the Department of Soil and Water Sciences for four years, from 1983 to 1987 and as a board member of the mapping of agricultural resources directed by the United Nations Development Programme (UNDP). He also served as an agricultural technical consultant to the Libyan African investment holding company. Dr. Azzeddin traveled to many countries in Africa on technical missions and to expose the soil and water resources in the African continent. He has contributed to education and research at the University of Tripoli through teaching research as one of the staff members at the Department of Soil and Water Sciences. He has been teaching graduate students since 1980 and supervised many theses and dissertations. Dr. Azzeddin has published several books, among which are Soil Survey, Classification and Land Evaluation (2015), Clay Minerals in Soil (2012), and Soil Profile Description in the Field (2009). Dr. Azzeddin helped to correlate the Russian soil classes to the American soil classification of Libyan soils. He also served as chairman of the technical team for surveying and classifying Libyan soils from 1980 to the present. As one of the technical team members, Dr. Azzeddin helped to prepare the soil map of Libya with a scale of 1:1 M (2005).

5 Major Limiting Factors Affecting Agricultural Use and Production

Basher A. Nwer, Khaled R. Ben Mahmoud, Hamdi A. Zurqani, and Mukhtar M. Elaalem

Abstract

Agricultural use and production in arid environments (i.e., Libya) can be affected by many numbers of constraints and limitations. Most of these constraints are related to the dry and hot climatic conditions and the absence of vegetative cover. These climatic conditions lead to the formation of soil types that have very poor soil attributes and limit their use for agricultural production. Soil limiting factors in Libya can be listed: unsuitable soil texture, soil depth, depth of water table, internal drainage, soil pH, exchangeable sodium percent, soil salinity, calcium carbonate contents, soil erosion by wind or water, and soil nutrients content. There are other constraints that can also be considered as the main limiting factors in agricultural productivity. They include climate, water resources, and terrain. Sustainable land resources management combines technologies, policies, and activities aimed at integrating socioeconomic principles with environmental concerns simultaneously, such as protecting and promoting sustainable use of terrestrial ecosystems, sustainably managing land use, combating degradation, and desertification. This chapter illustrates the constraints and limiting factors that negatively affect agricultural activities in Libya, which need to be addressed to improve the agricultural use and production in the country.

B. A. Nwer · K. R. Ben Mahmoud · H. A. Zurqani (✉)
M. M. Elaalem
Faculty of Agriculture, Department of Soil and Water Sciences, University of Tripoli, Tripoli, 13538, Libya
e-mail: h.zurqani@uot.edu.ly

B. A. Nwer
e-mail: bashir.nwer@yahoo.com

K. R. Ben Mahmoud
e-mail: benmahmoud.kh@gmail.com

M. M. Elaalem
e-mail: m.elaalem@uot.edu.ly

H. A. Zurqani (ed.), *The Soils of Libya*, World Soils Book Series,
https://doi.org/10.1007/978-3-030-66368-1_5

Keywords

Libya • Agricultural activities • Climatic • Topography • Soil limiting factors • Water scarcity

5.1 Introduction

According to the information provided in Chap. 4 about the properties of Libyan soils, some soil limiting factors that constraints agricultural use can be identified (Ben Mahmoud 1995, 2013). These factors can be summarized as follows: unsuitable soil texture, soil depth, depth of water table, internal drainage, soil pH, exchangeable sodium percent, soil salinity, $CaCO_3$ contents, soil erosion by wind or water, and soil nutrients content. According to Chap. 6, the area of cultivated land in Libya is estimated at about 210 thousand hectares (60 thousand hectares irrigated and 150,000 ha rainfed), which represents about 1.2% of the total area of the country. In addition to the previously mentioned soil limiting factors, the main determinant of the smallness of this area is the lack of water needed for agriculture in terms of quantity and quality (rain for rainfed crops or groundwater for irrigation). Consequently, soil, climate, water, terrain, and nature of crops are the factors that control the suitability of land for growing different crops. Therefore, these factors must be taken into consideration when determining the suitability of agricultural lands and add horizontally suitable areas for cultivation or increase the productivity of current agricultural lands. This chapter deals with the characteristics of Libyan lands in terms of their limitation to agricultural uses, which need to be addressed to improve the agricultural productivity in the country.

5.2 Soil Limiting Factors

As mentioned in Chaps. 3 and 4, the majority of areas in Libya are characterized by dry and hot climatic conditions and the absence of vegetative cover. This leads mainly to the formation of Entisols, Aridisols. Mollisols, Alfisols, and Vertisols, which are only found in Al-Jabal Al-Akhdar, Inceptisols are also found in the Al-Jabal Al-Akhdar and in a small area of the Tripoli Mountains (Jabal Nafusa), where the Mediterranean climate is dominating. Such soils have a low content of organic matter (O.M %) and low nutrient content. In addition, soil physical attributes, for example, soil texture, infiltration rates, cannot support plant production well (Nwer 2005). There are many soil limiting factors in the agriculture system. Available water is essential for plant growth. The water storage capacity of the soil is one of the most important soil properties for soil functions including productivity (Shaxson 2006; Jones et al. 2009). The other group of limitations include other internal soil deficiencies mainly due to an improper substratum, limiting rooting and nutrition of plants. These include shallow soils, stoniness, hardpans, anaerobic horizons, or soils with adverse chemistry such as salinity, sodicity, alkalinity, nutrient depletion, or contamination, which may cause severe restrictions to plant growth or utilization of biomass (Murray et al. 1983; Louwagie et al. 2009; Ben Mahmoud 2013). Other soil constraints to agricultural use are the depth of the water table, internal drainage, and soil erosion by wind or water (Ben Mahmoud 1995). The following sections will address soil properties that are believed to have a greater impact on agriculture production in the Libyan context. According to Ben Mahmoud (2013), these problems and obstacles can be summarized as follows.

5.2.1 Soil Texture

Texture is the most permanent characteristic of soil. It decisively influences a number of soil attributes such as soil moisture regime, permeability, infiltration rate, runoff rate, erodibility, workability, root penetration, and fertility (ILACO 1989). Soil texture is the most important property to be determined, and it is critical for understanding soil behavior and management. From the soil texture, many conclusions can be drawn. Soil texture governs moisture and nutrient storage capacity. It provides a measure for permeability, and to some extent, for water retention capacity (Brady and Weil 1999). With the exception of some Libyan soils that have relatively suitable texture, the majority of these soils are Coarse sandy soils, compact clay soils, and gravel and stony soils that are unsuitable for agricultural use.

Fig. 5.1 Stony gravel soils (Ben Mahmoud 2013)

This is because they reduce the chance of crop growth, cause difficulties in agricultural activities, and have negative effects on soil-water properties (Fig. 5.1).

5.2.2 Soil Depth

Soil depth can greatly influence the types of plants that can grow. Soil depth is a very important parameter in determining the quantity of water and nutrients that can be stored in the soil profile, as well as the ability to support plant growth, during periods of no rainfall or when irrigation ceases. Soil depth is also important in determining the ability of the profile to support plant roots and the ability of the crop to withstand wind damage. Under irrigated conditions, soil depth affects drainage, aeration, and water retention properties. The available storage capacity of the soil and consequently the water application depth and irrigation frequency are determined by root depth and root distribution, as well as by the moisture characteristics of the soil (ILACO 1989; Nwer 2005). Deeper soils can provide more water and nutrients to plants than more shallow soils. Furthermore, most plants rely on soil for mechanical support. Deep soils favor drainage and are, therefore, optimal for the irrigation of dryland crops. Stewart and Nielsen (1990) indicate that a soil depth of about 30–40 cm is the minimum depth for maximum yields of all crops. Significant decreases in production are evident for all crops where soil depth is less than 30 cm. The shallow marginal soils only support certain plants, whereas deep soils can support a wide range of plants and trees.

In Libyan conditions, it has been suggested that a soil depth of less than 30 cm vastly decreases the crop yield. In addition, it is agreed that a soil depth between 50 and 80 cm increases the crop yield (Selkhozpromexport 1980; Stewart and Nielsen 1990; Ben Mahmoud 1995; Food and Agriculture Organization of the United Nations (FAO) 2002; Nwer 2005). Libyan soils in semiarid and arid regions can develop special soil horizons that limit the soil depth available to support the plant. These special horizons include hardpans, duripans, and layers that have been cemented together by the deposition of mineral material. Petrocalcic and Petrogypsic horizons are obvious examples of this kind of horizon. Aridisols and Entisols represent the most widespread soils in Libya (see Chaps. 3 and 4). There are a large number of Libyan soils that are <50 cm in depth, and consequently, most of these soils are not suitable for irrigation. Shallow soils lack the depth necessary to stabilize the roots of trees and other crops, making agriculture and irrigation very difficult. Additionally, poor organization of irrigation processes in these soils may lead to an increase in soil salinity and waterlogging (Fig. 5.2).

Fig. 5.2 Shallow soils (Ben Mahmoud 2013)

5.2.3 Depth of Water Table

The depth of the water table is considered one of the basic factors in the soil that impede the growth of agricultural crops and reduce its productivity, as the presence of water table close to the surface affects mainly the aeration of the soil, and thus the quality of the microbes that live in it. It also affects the internal soil drainage. Many southern depression areas (general agricultural development projects in the Libyan Desert) suffer from this phenomenon if these lands are placed under irrigation. This phenomenon appears naturally in Libyan soils such as Salids and Gypsids in many areas in Libya (west, middle, and south) and in some small, low areas in Jabal Al Akhdar (local terrain) in which Natrixeralfs are found (Ben Mahmoud 1995).

5.2.4 Soil Salinity

Soil salinity is a serious problem in arid and semiarid zones of the world, where the poor quality of water is often the only source available for irrigation. Salts tend to accumulate in the upper soil profile, especially when intense evapotranspiration is associated with insufficient leaching. The addition of salts to the soil alters its physical and chemical properties, including soil structure and hydraulic conductivity (Tanji 1996). Salinity affects plants by inhibiting the uptake of water by osmosis. Moderate salinity levels retard growth and reduce yield, while high levels kill crops and may cause areas to be barren of plants. Salinity can affect plant growth by reducing the amount of water available to the crop and by increasing the concentration of certain ions that have a toxic effect on plant metabolism (FAO 1995; Nwer et al. 2013). In 1980, Selkhozpromexport reported that the Libyan lands covered by saline soils or soil affected by salts reached 200 thousand hectares in the northwestern region (12% of the total area of this region). While, in the northeastern region, saline soils reach about 334 thousand hectares (22% of the total area of this region) (Ben Mahmoud 1995).

Mapping and monitoring salinity is vital to keep track of and anticipate further degradation and is essential for proper and timely interventions to adjust management practices or undertake suitable reclamation and rehabilitation measures (Zurqani et al. 2019). Mapping and monitoring salinity involves first identifying the areas where slats concentrate;

second, it requires detecting the temporal and spatial changes in this occurrence. Ben Mahmoud et al. (2000) reported that approximately 1900 km^2 are affected by salinity, probably due to using saline water for irrigation, poor drainage, and increasing concentrations of salts in the irrigated water from seawater intrusion. There are many reasons for the salinity problem in Libyan soils including: sequences of arid years, low quality of irrigation water, high evaporation rate from the soils due to the high temperature, bad drainage of the soils, soil saturation, and interruption between groundwater and salty seawater. In addition to the presence and spread of saline and natural Sebkha soils, significant areas of irrigated lands in Libya are annually transformed into saline soils. Secondary salinity may also arise in some Libyan soils (Ben Mahmoud et al. 2000) (Fig. 5.3a, b).

5.2.5 Soil Sodicity

Sodic soils are those which have an exchangeable sodium percentage (ESP) of more than 15 meq/100 g and pH more than 8.5 (Ben Mahmoud 1995). Excess exchangeable sodium has an adverse effect on the physical and nutritional properties of the soil, with a consequent reduction in crop growth, either significantly or entirely (Zurqani et al. 2018). Sodic soils can be formed naturally (affected by the special condition of soil forming factors) such as the case of the presence of the Natrargids of southwest Benghazi city, or when a clayey soil is put under irrigation with watercontaining high soluble sodium. In late cases, the exchangeable calcium ions can be displaced and replaced by sodium ions and sodic soils can be formed.

a

b

Fig. 5.3 **a** Coastal Sebkha (Ben Mahmoud 2013). **b** Secondary Salinization (Ben Mahmoud 2013)

In sodic soils, sodium ions are attached to clay particles. The monovalency of sodium cations leads to a loss in the stickiness of clay particles when wet. If the proportion of divalent ions is reduced, the aggregate formation will also be reduced. This leads to collapsed and unstable soils that become impermeable to water and roots and which erode easily (Tanji 1996; ILACO 1989; Landon 1984). The FAO (1988) conclude that plant growth is adversely affected in sodic soils due to one or more of the following factors: firstly, high ESP in sodic soils markedly influences the physical soil properties. Secondly, the effect of ESP on plant growth is through its effect on soil pH. A high pH could affect greatly lowering the availability of essential plant nutrients such as Calcium, Magnesium, phosphorus, iron, manganese, and zinc.

Yeha (1982) classified the plants in Libya into five categories according to ESP tolerance. The classification was based on the USDA classification. Ben Mahmoud (1995) points out that ESP in some Entisols in the south of Libya has been found to be between 3 and 28%. Ben Mahmoud believes that the main reason for this high ESP is not the high sodium but the low CEC in these sandy soils. He argues that the CEC is very low and ranges from 1 to 10 me/100 g soil. Therefore, crops in such soils may not be affected by the sodium even though the ESP appears to be high. This could be true in the south of the country and some parts of the North West; however, in the North East of Libya, the CEC is relatively higher (Zurqani et al. 2018).

In general, soil sodicity is believed to be one of the limiting factors facing agriculture development in Libya. The effect of soil sodicity is apparent in Libya, especially in the North East of the country where the good soils exist with high clay content and relatively high soil cation exchange capacity (CEC).

5.2.6 Calcium Carbonate Content

Carbonates in soil profiles may be derived from carbonate-rich rocks but it can be encountered as secondary depositions from groundwater (Landon 1984). Calcium carbonate ($CaCO_3$) in the form of free lime in the soil profile affects soil structure and interferes with infiltration and the evapotranspiration process. It influences both the soil moisture regime and the availability of nutrients (FAO 2002). The higher percentage of calcium carbonate in the soil, the worse its various properties, and consequently the lower its productivity and the fewer the chances of cultivating crops in it. The presence of calcium carbonate in high content leads to weak soil structure and thus increases the degree of water loss throughout the soil profile. The high percentage of calcium carbonate leads to a decrease in the CEC and a rise in the pH. And the last one leads to the conversion of some nutrient elements such as phosphorous, iron, and zinc into a form that is not available for absorption by plants.

Calcium carbonate being only slowly soluble does not affect plants at the percentage levels at which soluble salts become harmful but may have adverse effects when present in high concentrations, including as calcic horizons. This is most likely to occur in semiarid regions (ILACO 1989). Calcium carbonate can form hardpans, including Petrocalcic horizon, in which calcium carbonate cement the soil particles. The net effect of these dense horizons is to delay or prevent root growth and thus limit the effective depth of the soil. They also may affect soil oxygenation by restricting drainage at times in which large amounts of water are present. Therefore, calcium carbonate represents one of the limiting factors of agricultural production. In Libya, the dry climatic conditions help to have a high accumulation of calcium carbonate in the soils. Calcids, Petrocalcids, and Rendolls are the major soils that have a high percentage of $CaCO_3$ (Ben Mahmoud 1995).

5.2.7 Nutrients

Libyan soils are characterized by their low content of nitrogen, available phosphorous, and available trace elements, especially zinc and iron. Nitrogen is bound in the soil with organic matter, which is considered low to very low in Libyan soils. As for the lack of available phosphorus in the Libyan soils, it is mainly due to the increase in soil pH and the content of calcium carbonate, which turns it into a form that is not available for absorption by plants. Also, the high soil pH level decreases the availability of zinc, iron, and in some cases manganese and copper. Therefore, if these nutrients are not added during the cultivation of agricultural crops in Libyan soils in the appropriate quantity and manner, then these elements are considered a factor limiting the growth of most agricultural crops and reducing the productivity of some others (Ben Mahmoud 1995).

5.3 Other Constraints to Agricultural Use and Development

There are many other constraints to agriculture use and development in Libya other than the soil factors such as climate, water resources, terrain, and soil erosion. These factors can be also considered as the main limiting factors in agriculture productivity. The following is a brief description of these factors and their influences on agricultural expansion and increasing agricultural productivity.

5.3.1 Climate

The climate controls the necessary needs for the growth and production of various agricultural crops, the most important of which in Libya are the light, thermal, and water needs. In addition to the damage, related to agricultural crops, caused by the winds, especially the hot and dry ones. The Libyan climate has already been reviewed in Chap. 3. In the following, there is a special mention of the climatic factors in terms of their constraint on the growth and production of agricultural crops. Rain is considered one of the important climatic factors in areas that depend on rainwater for their cultivation as the only source of water (such as the rainfed areas in the Jafara Plain, the Tripoli Mountains, Al Jabal Al Akhdar, and some northern wadis). The prevailing temperatures in a region are also one of the most important factors determinants the growth and production of different agricultural crops. These crops vary among themselves in their ability to withstand temperatures and the extent of their change. Therefore, only certain crops are able to grow and produce under different thermal conditions in Libya. Light as a climatic factor has an essential role in the formation of chlorophyll, and in the processes of photosynthesis. In general, the length of the day for agricultural crops is often the limiting factor for any elongation or shortening of phalanges, and for the formation or non-formation of flowers.

Moreover, plants in light temperate regions (such as in Libya), grow either acclimatized for a long day or short day (summer crops compared with winter crops). As a result of the geographical location of Libya and its presence in areas in the temperate regions in terms of light, light is not considered a determining factor for agricultural growth and production.

As for the component of the wind, as previously explained in Chap. 3, winds of different types and directions may blow over Libya. Despite the positive effect of these winds in the formation of clouds, air purification, and the transfer of pollen and others, the strong winds, especially the hot dry ones, lead to some physiological, anatomical, and mechanical damage to agricultural crops, and the wind becomes one of the main factors that hinder agricultural crops in places and time its activity.

Relative humidity also affects agricultural crops due to its effect on the amount and speed of water that is lost through transpiration and evaporation, the higher the rates of transpiration and evaporation, the higher the water requirements of crops. From the information contained in Chap. 3, the relative humidity is high in the northern regions of Libya and gradually decreases towards the south (The relative humidity in Jabal Al Akhdar reaches more than 80%, while it reaches <30% in Al Kufra). Thus, it can be concluded that the water requirements for a specific crop such as wheat are much more when it is grown in Al Kufra than when it is grown in the Al Marj area of the Jabal Al Akhdar (Ben Mahmoud 1995).

5.3.2 Water Resources

Water is the main constraint to agricultural development in Libya. Water quantity and quality are the main issues related to the usage of agriculture. Groundwater is the main water source in Libya. It accounts for more than 88% of the total water consumption (Salem 1992). The total volume of groundwater in Libya is estimated to be 99,500 km^3 with an uncertainty range of between 64,600 and 234,000 km^3 (Fig. 5.4) (MacDonal et al. 2012).

It is clear from Fig. 5.4 that groundwater is the main source of water for agriculture usage. The majority of this resource is nonrenewable; thus, in the long term, agriculture cannot be based on such a resource. Quality and quantity issues also risk such resources. In the north of the country, seawater intrusion has affected the coastal areas, whereas in the south, groundwater resources are not renewable (Ben Mahmoud 2013). Salem (2007) argued that even if 50% of the water can be intercepted from rainfall and form a resource, this 100 million m^3 /yr will constitute only 3% of water resources. It can be concluded that water resources mainly depend upon groundwater, and other water resources are yet to make a considerable contribution to water needs and demands. This has put a serving pressure on groundwater resources in coastal regions and deserts. This pressure led to water ground drawdown and seawater intrusion. In the coastal area, groundwater quality has been badly affected.

The main obstacle to agriculture development is water resources. Therefore, there is a pressing need to develop alternative agriculture practices taking this limitation into

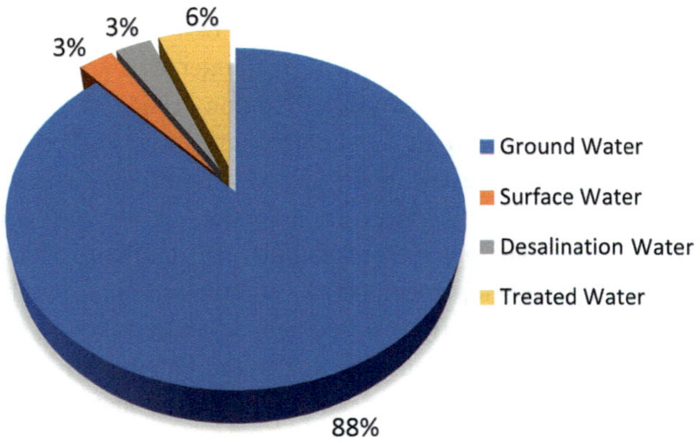

Fig. 5.4 Water resources in Libya adapted from (Salem 2007)

consideration. This can be done by targeting crops that consume less water, and at the same time, developing water resources other than groundwater resources.

5.3.3 Terrain

Since the topography plays a fundamental role in choosing agricultural crops, because of their impact on the local climate, vegetation cover, the type of the existing soil, the local humidity and temperature conditions, the extent of their exposure to erosion by wind or water, and others. However, what concerns us here is one of its elements that controls the methods that must be followed when growing agricultural crops, especially irrigated ones, which is the slope of the land surface. The slope of the land surface is one of the important characteristics that control the choice of agricultural crops, as well as the selection of agricultural operations that must be followed. Such as the use of agricultural mechanization or not, the type of irrigation system, measures related to soil and water conservation, etc.

5.3.4 Erosion

Soil erosion leads to a reduction in soil quality, productivity, and crop yield. Its impact extends to several variables such as crop type, soil properties, management practices, and climate characteristics. Erosion often results in a decrease of the soil supply functions in three ways: (1) the removal of organic matter; (2) the change in depth to a possible root-barrier; and (3) the loss of structure and increased compaction (Bakker et al. 2004). Erosion may also cause physical hindrance to root growth, for example, when a clayey subsoil is present. Physical hindrance is a significant part of the root system encountering the restricting horizon. Where growth is hindered by bedrock or a pan, yields will rapidly approach zero once the minimum soil requirements for rooting are exceeded by soil removal (National Soil Erosion—Soil Productivity Research Planning Committee 1981).

In Libya, where the environment is vulnerable, the variability of rainfall, the occurrence of occasional relatively heavy showers characterized by high intensity, and the land surface are the main factors that accelerate soil erosion. The combination of these factors, in addition to the topography, has increased the rate of soil erosion by water. Selkhozpromexport (1980) distinguished two types of accelerated erosion: Soil erosion by wind and Soil erosion by water.

- **Soil Erosion by wind**: The majority of Libyan soils are exposed to erosion by wind and that is due to the arid climate, the absence of vegetation cover, the sandy soil texture, the dominant high wind speeds, and the unsustainable land uses (Fig. 5.5).

This phenomenon is one of the most prevalent environmental problems in Libya, whether in terms of the loss of fertile topsoil or the problems caused by the movement,

Fig. 5.5 Soil erosion by wind (Ben Mahmoud 2013)

Fig. 5.6 Soil erosion by water (Ben Mahmoud 2013)

encroachment, and fall of sand on residential areas. This can lead to things like the disruption of transportation routes and the spread of eye and respiratory diseases.

- **Soil Erosion by water**: The Northern mountain regions in Libya are the most vulnerable areas for soil erosion by water. This is due to stormy rains, the absence of vegetation cover, variability in the topography (slopes), low capacity to hold water, shallow soil profiles, low content of organic matter, and unsustainable land uses (Fig. 5.6).

Soil erosion by water leads to the loss of the fertile surface layer, and the transfer of large amounts of soil to other sites, leaving behind rocky exposures, shallow soil, or dead lands. These eroded materials then collect in areas behind dams or in agricultural fields, causing damage to the land and further damaging agriculture.

Soil erosion by water is common in the form of Sheet or Gully, occurring mainly within the Al-Jabal Al-Akhdar and Jable Nafusa Uplands, while soil erosion by wind is found in the form of deflation within the littoral plain (Selkhozpromexport 1980; Ben Mahmoud 1995; Nwer 2005). Table 5.1 shows the size of the problem in Libya, and especially in the northwest and northeast regions.

Soil erosion can cause an obstacle to agricultural development and production in Libya. It affected thousands of hectares of soil. This can greatly impact efforts to soil rehabilitation and combat soil erosion, where the cost of rehabilitation proved to be high in most cases (Zurqani et al. 2019).

5.4 Soil Management and Future Needs

Soil makes up the fundamental resource-base in any agricultural production system, and hence, its appropriate management is vital for sustainable agricultural production.

Nowadays, countries must achieve an increase in production to be successful in agriculture. This can only be achieved by implementing sustainable methods and solutions in agriculture. The fact that the agricultural activities and practices are compatible with the environment and are permanent has great importance in terms of contributing to the sustainability of the ecology (Pretty 2008).

According to the FESLM: An International Framework for Evaluating Sustainable Land Management definitions by

Table 5.1 Soil erosion by water in Libya

Erosion type	Area (1000 ha)	
	North West region	North East region
Sheet erosion		
• Slight	155.5	241.7
• Moderate	154.5	41.7
• Severe	54.5	1.7
Gully erosion		
• Slight	85.3	0.8
• Moderate	57	0.0
• Severe	511	285.7

Source Selkhozpromexport (1980), Ben Mahmoud (1995), Nwer 2005)

FAO, sustainable land management combines socioeconomic principles with environmentally sensitive technologies, policies, and activities. In order for sustainable land management to be feasible, five objectives have been identified as Efficiency, Security, Protection, Vitality, and Acceptability. The implementation and findings of the SLM regulation have been identified as the main pillars to be tested and monitored. In view of this framework, the SLM regulation should be followed cautiously in the field and established in a production model that will protect the natural resources. In addition, the model should be economically feasible and socially acceptable. However, it should also be understood that the system cannot be sustainable in practices where the agricultural structure is not properly managed, and the land is constantly destroyed. This method requires, in principle, to protect and improve soil fertility, to prevent and correct soil degradation, and to prevent environmental damage.

In summary, sustainable soil resources management combines technologies, policies, and activities aimed at integrating socioeconomic principles with environmental concerns simultaneously, such as protecting and promoting sustainable use of terrestrial ecosystems, sustainably managing land use, combating degradation and desertification, etc.

5.5 Conclusions

In Libya, agricultural activities are highly sensitive to the limitations that are related to the natural environment as a result of the unique climate, topography, soil characteristics. Also, water scarcity and the low quality of water available are among the common obstacles to agriculture production in the country. Hence, farmers and farm communities across the country are continuously challenged by these conditions. To be successful in agricultural use in Libya, it is crucial that the mentioned constraints and limitations are taken into account in future plans. In fact, agricultural sustainability can be only achieved if sustainable land management combines socioeconomic principles with environmentally sensitive technologies, policies, and activities. An increase in agricultural production in Libya can be achieved by taking some measures such as the adaptation of agro-components, for example, fertilization, and the maintenance, mechanization, and harvesting of suitable land. In addition, the application of best soil management practices and combating desertification and land degradation.

References

Bakker MM, Govers G, Rounsevell MDA (2004) The crop productivity-erosion relationship: an analysis based on experimental work. CATENA 57:55–76

Ben Mahmoud KR (1995) Libyan soils (Their genesis, classification, properties and agricultural potentials). NASR, Tripoli, Libya, pp 615 (in Arabic)

Ben Mahmoud KR (2013) Towards a national strategy for the sustainability of natural resources and enhancing food security in Libya, pp 123 (in Arabic)

Ben Mahmoud KR, Mansur S, L-Gomati A (2000) Desertification in the third millennium, desertification conference toward better management of Arid and Semi-Arid Lands in the 21st Century, Dubai, United Arab Emirates. Zayed Prize Fdn, United Arab Emirates. https://www.routledge.com/Desertification-in-the-Third-Millennium-Proceedings-of-an-International/Alsharhan-Wood-Goudie-Fowler-Abdellatif/p/book/9789058095718. Accessed 5 Aug 2020

Brady NC, Weil RR (1999) The nature and properties of soil, 12th edn. Mac. Pub. Com. New York, pp 625–640

Food and Agriculture Organization of the United Nations (FAO) (1988) Salt-Affected Soils and their Management. http://www.fao.org/3/x5871e/x5871e00.htm. Accessed 5 Aug 2020

Food and Agriculture Organization of the United Nations (FAO) (1995) Sustainable dryland cropping in relation to soil productivity. http://www.fao.org/3/v9926e/v9926e00.htm. Accessed 5 Aug 2020

Food and Agriculture Organization of the United Nations (FAO) (2002) Agricultural drainage water management in arid and semi-arid areas. http://www.fao.org/3/a-ap103e.pdf. Accessed 5 Aug 2020

ILACO (1989) Agricultural compendium for rural development in the tropics and subtropics, 2nd edn. Elsevier Science Publishing Company Inc., New York. https://library.wur.nl/WebQuery/titel/501212. Accessed 5 Aug 2020

Jones A, Stolbovoy V, Rusco E, Gentile AR, Gardi C, Marechal B, Montanarella L (2009) Climate change in Europe. 2. Impact on soil. A review. Agron Sustain Dev 29(3):423–432

Landon JR (1984) Booker tropical soil manual. Booker Agriculture International Limited, New York

Louwagie G, Gay SH, Burrell A (2009) Addressing soil degradation in EU agriculture: relevant processes, practices and policies. Report on the project 'Sustainable Agriculture and Soil Conservation (SoCo). JRC Scientific and Technical Reports, ISSN 1018 5593, 209 p

MacDonal AM, Bonsor HC, Dochartaigh BEO, Taylor RG (2012) Quantitative maps of groundwater resources in Africa. Environ, Res Lett, p 7024009

Murray WG, Harris DG, Miller GA, Thompson NS (1983) Farm appraisal and valuation, 6th edn. Iowa State University Press, 304 p

National Soil Erosion-Soil Productivity Research Planning Committee (1981) Soil erosion effects on soil productivity: a research perspective. J Soil Water Conserv 36(2):82–90

Nwer B (2005) The application of land evaluation technique in the north-east of Libya. Ph.D. dissertation, Faculty of Environment, Cranfield University, Silsoe

Nwer B, Ali A, Zurqani H (2013) Soil salinity mapping model developed using RS and GIS in Libya. In: Global science & technology forum. http://www.geosp.net/?research=soil-salinity-mapping-model-developed-using-rs-and-gis-in-libya

Pretty J (2008) Agricultural sustainability: concepts, principles and evidence. Philosophical Trans Royal Soc B 363:447–465
Salem OM (1992) The Great Manmade River Project. GMRP, Tripoli, Libya
Salem O (2007) Management of shared groundwater basins in Libya. African Water J 1(1):106–117. www.amcow-online.org/docs/journal/African%20Water%20Journal%20-%202007.pdf. Accessed 5 Aug 2020
Selkhozpromexport (1980) Soil studies in the Eastern zone of the Socialist People's Libyan Arab Jamahiriya. Secretariat for Agricultural Reclamation and Land Development, Tripoli
Shaxson TS (2006) Re-thinking the conservation of carbon, water and soil: a different perspective. Agron Sustain Dev 26:9–19. https://open-furrow-agrosphere.net/Documents/UTW/Shaxson-2006-HAL.pdf. Accessed 5 Aug 2020
Stewart BA, Nielsen DR (1990) Irrigation of agricultural crops. Agronomy Series No. 30. Madison, USA
Tanji KK (1996) Agricultural salinity assessment, and management. ASCE Manuals and Reports on Engineering No 71, pp 113–161
Yeha A (1982) Management and reclamations of saline soils in Libya. AL-Fateh University, Tripoli
Zurqani HA, Mikhailova EA, Post CJ, Schlautman MA, Sharp JL (2018) Predicting the classes and distribution of salt-affected soils in Northwest Libya. Commun Soil Sci Plant Anal 49(6)
Zurqani HA, Mikhailova EA, Post CJ, Schlautman MA, Elhawej AR (2019) A review of libyan soil databases for use within an ecosystem services framework. Land 8(5):82

Dr. Bashir A. Nwer was born in Libya in 1971. He is a Ph.D. graduate from Cranfield University, UK, 2006, in the field of land evaluation using geographic information System (GIS). Dr. Nwer has great interest in the field of land degradation and combating desertification. He was Chairman of the National Committee to Combat Desertification in Libya. He was vice-chairman of the committee of the review of the implementation of UNCCD and a rapporteur of the Ninth and Tenth Session of CRIC, UNCCD. Dr. Nwer has attended UNCCD, COP5-COP11, and most of the regional meetings of UNCCD. Dr. Nwer has published a number of papers in the use of GIS and RS in land evaluation and degradation.

Dr. Khaled R. Ben Mahmoud is a senior professor of soil pedology. He is a distinguished soil scientist with wide range of scientific and working experiences in Libya and abroad. He received his Ph.D. in 1977 from Colorado State University (USA). His major teaching activities and research are at the Department of Soil and Water, Faculty of Agriculture, University of Tripoli, Libya. He has been teaching undergraduate and graduate students in Libya and has supervised many M.Sc. and Ph.D. thesis and dissertation in Libya and abroad. He focused mainly on soil resources, land evaluation, and mapping. He contributes to soil science in Libya and abroad as the Head of the Department of Soil and Water Sciences, Faculty of Agriculture, University of Tripoli, Tripoli, Libya (1978–1983), Dean of the Faculty of Agriculture, University of Tripoli, Tripoli, Libya (1983–1986), Chairman of the Committee for Libyan Agricultural Mapping, Tripoli, Libya (1988–1990). A Research Coordinator, University of Tripoli, Tripoli, Libya (1997–2001), Member of the Scientific Council of Agricultural Research Center (ARC), Tripoli, Libya (2000–2002), Member of the Scientific Council of the College of Higher Studies at the University of Tripoli, Tripoli, Libya (2000–2002), Member of the Scientific Committee of Libya Centre for Remote Sensing and Space Sciences, Tripoli, Libya (1998–2006), A National Coordinator of the Project implemented by FAO: Mapping of Natural Resources for Agricultural Use and Planning Tripoli, Libya (2000–2006), A General Coordinator of the National Programme for the Development of Vegetation Cover, Tripoli, Libya (2004–2006). Director of the Division of Land and Water Uses, The Arab Centre of the Studies of Arid Zones Lands (ACSAD), The League of Arab States, Damascus, Syria (2007–2011), Head of the Programme: Monitoring and Combating of Desertification in the Arab World, The Arab Centre of the Studies of Arid Zones Lands (ACSAD), The League of Arab States, Damascus, Syria (2007–2011), Director General of Eddayra Co. (Agricultural and Environmental Consulting and Services), Tripoli, Libya (2011–to date). Engaged as a local and international expert to control, evaluate, approve, and accept of many soils studies and mapping which have been implemented by different foreign and local companies, agencies, and offices in different parts of Libya and abroad (1977–to date). More than 100 country reports, research and scientific papers, and other publications were published in different local and international scientific journals and symposiums (in Arabic and English). He is the author of more than 10 books that have been written in Arabic. The most important ones are: Sandy soils of Libya (1981), The study of soils in the field (1984), Principles of Soil Science for Agricultural Secondary and Technical Schools (1985), Fundamentals of Soil Science and its Relation to Plant Growth (1995), Libyan Soils (Their Genesis, Classification, Properties, and Agricultural Potentiality (1995), Fundamentals of Soil and water Science (2001), Issues on Desertification and Land Degradation in the Arab Region (2010), Towards a National Strategy for the Sustainability of Natural Resources and Enhancing Food Security in Libya (2013), and The Libyan Desert (its natural potentials and prospects for investment) in press. Finally, he was appointed as a technical advisor to the President of the Libyan Presidential Council for Agriculture, Tripoli, Libya (2018–to date).

Dr. Hamdi A. Zurqani is one of the faculty members at the Department of Soil and Water Sciences, Faculty of Agriculture, University of Tripoli, Tripoli, Libya. Dr. Zurqani is a recognized expert as a result of his internationally acclaimed work in the areas of environmental information science, remote sensing, land evaluation, sustainability, pedology, and soil science education. He has conducted research across the world, including the United States of America and Africa. Dr. Zurqani is a distinguished soil scientist with a wide range of scientific and working experiences in Libya and abroad. He received his M.Sc. (2010) from the University of Tripoli, Tripoli, Libya and Ph.D. (2019) from Clemson University, Clemson, SC, USA. His major research and teaching activities at the University of Tripoli have focused mainly on soil genesis and classification and the environmental information sciences (Remote Sensing and Geographic Information System). He has published broadly in many journals (e.g., Nature "Scientific Reports", Geoderma; International Journal of Applied Earth Observation and Geoinformation; Journal of Geological Sciences; Land; Frontiers in Environmental Science; Communications in Soil Science and Plant Analysis; and others). Dr. Zurqani is a member of the Editorial Board for Remote Sensing (MDPI) Journal, counseling outcome and research evaluation. He also was appointed to serve as a Guest Editor for the Special Issue "Applications of Remote Sensing in Earth Observation and Geo-Information Science". In addition, Dr. Zurqani conducted peer-review for many journals including Journal of Environmental Informatics, Applied Sciences, SN Applied Sciences, Remote Sensing, Heliyon, Geosciences, Land, Water, Agronomy, Agriculture, Sustainability, Arid Land Research and Management, International Journal of Environmental Research and Public Health, Natural Hazards, and Conference of the Arabian Journal of Geosciences. He is also one of the authors of the lab manual entitled, "GIS Exercises for Natural Resource Management". Dr. Zurqani has been the recipient of numerous awards and honors: Recipient of Douglas R. Phillips Award for Graduate Students, Department of Forestry and Environmental Conservation, Clemson University, April 12, 2019; the First Place Best Judged Poster (CAFLS) at the GRADS 2019: Clemson Student Research Forum on April 4, 2019; the Second Place Poster at the 11th Clemson Biological Sciences Annual Student Symposium, April 6, 2019; the Second Place Best Judged Poster at the Clemson Student Research Forum on April 4, 2018; and the Third Place Poster at the 9th Clemson Biological Sciences Annual Student Symposium, February 25, 2017. Dr. Zurqani conducts cutting-edge research in the field of environmental information science, remote sensing, land use management/planning, change detection of landscape degradation, and geographic information system (GIS) models. He has focused on his research efforts on the development of new technologies in the field of environmental information sciences, geo-intelligence (advanced geo-information science and earth observation, machine and deep learning, and big data analytics), remote sensing, land evaluation, pedology, land use management/planning, monitoring and evaluating sustainable land management, change detection of landscape degradation, and geographic information system models.

Dr. Mukhtar M. Elaalem is an associate professor of soil pedology and head of the Department of Soil and Water Sciences, Faculty of Agriculture, University of Tripoli, Tripoli, Libya. He got a Ph.D. degree in Geographic Information System at Leicester University, UK in 2011. His Ph.D. thesis is on "The Application of Land Evaluation Techniques in Jafara Plain in Libya using Fuzzy Methods", and his research interest is soil pedology, geographical information science, natural resources applications of GIS (especially land evaluation and soils studies), modeling spatial uncertainty, especially employing fuzzy logic methods to land evaluation and soils studies. He teaches various courses, including introduction to soil science, soil survey; genesis and classification, and Libyan soils. He attended many local, regional, and international meetings, workshops, training programs, conferences on land resources, and applications of GIS and remote sensing.

6 Land Cover, Land Use, and Vegetation Distribution

Hamdi A. Zurqani and Khaled R. Ben Mahmoud

Abstract

Vegetation refers to the plants and trees that are found in a particular region. The geographical location and climatic conditions of Libya strongly influence the vegetation status. The diversity of natural vegetation in Libya varies according to the climatic zone and soil types. Acknowledging that most of the country consists of desert dry areas, the fertile regions are mainly present in the coastal zone (including plains and mountains) and scattered depressions and Wadis in the desert. In 2009, the Food and Agriculture Organization (FAO) (LIB/00/004 "Mapping of Natural Resources for Agriculture Use and Planning in Libya Project") published the first land cover map of the country using Remote Sensing (RS), Geographic Information System (GIS) techniques, GeoVIS Software and Land Cover Classification System (LCCS) which was approved by the FAO. This map shows the main categories of land cover types according to location, quantity, and quality of the natural vegetation cover. This map, as well as its accompanying statistics, is considered one of the most reliable sources of the land cover in Libya. The agricultural land is estimated at about 8.73% of the total area of the country with an arable land area of about 1%. Crop productions essentially depend on the private sector, where most of the farms are relatively small areas with grains and some vegetables, fruits, and forage crops. In this chapter, we summarize the existing information about land cover, land use, and vegetation distribution in Libya.

Keywords

Libya • Natural vegetation • Agriculture • Crop productions • Cropping pattern • Suitability

6.1 Introduction

The vegetation cover is considered one of the most important factors in soil formation. Vegetation cover has a substantial impact on the stabilization of soil structure and the protection of soil erosion (Mahmoudzadeh 2007). It also improves some of the soil's physical-chemical properties through the physical protection of the soil surface, by the contribution of organic matter from both the plant canopy and the root system (Ruiz-Colmenero et al. 2013). Plant diversity affects both the quantity and quality of plant-derived residues entering the soil subsystem (Chen et al. 2016). Usually, the vegetation cover is divided into three different vegetation types, which are grassland or pasture, shrubs, and trees (including forests). These vegetation types are mainly distributed in the country along a north to south gradient of precipitation and according to different soil types, as well as precipitation variability in an east to west direction. Climatic conditions, physiographic regions, varied topography, and the influence of anthropogenic activities play an essential role in the diversity of the vegetation cover in Libya (Zurqani et al. 2019). As most of the Libyan land is desert, the distribution of the natural vegetation has a very low density, consisting of either separate pygmy desert shrubs or desert herbs, and it can often be found near the oases (Ben Mahmoud 1995).

In 2001, the statistical office of the European Union (EUROSTAT) defined the land cover as the observed physical cover of the earth's surface, while it labeled land use as the description of the socioeconomic function of the same area (EUROSTAT 2001). The land cover density is strictly dependent on rainfall patterns, particularly in arid

H. A. Zurqani (✉) · K. R. Ben Mahmoud
Department of Soil and Water Sciences, Faculty of Agriculture, University of Tripoli, Tripoli, 13538, Libya
e-mail: h.zurqani@uot.edu.ly

K. R. Ben Mahmoud
e-mail: benmahmoud.kh@gmail.com; ; m.elaalem@uot.edu.ly

H. A. Zurqani (ed.), *The Soils of Libya*, World Soils Book Series,
https://doi.org/10.1007/978-3-030-66368-1_6

zones (Maselli et al. 2009). Libya is characterized by similar vegetation types that closely resemble the quality and density of the species found in climatically similar regions (dry desert areas and Mediterranean regions), with the exception of coastal areas, and especially the northeastern and northwestern mountains (the mountains of Tripoli and Al Jabal Al Akhdar), which receive rainfall of more than 200 mm annually (Al-Idrissi et al. 1996). Most of the agricultural activities are restricted to a long narrow strip along the Mediterranean coast, low mountains, and scattered oases in the desert (Zurqani et al. 2019) (see Fig. 1.3, Chap. 1).

Previously, there were no accurate statistics on the types of vegetation cover in Libya due to the diffuse boundaries between vegetation and arable land. However, this problem has been resolved in the recent land cover map (LIB/00/004) in terms of species and the most predominant vegetation types. This chapter discusses the land cover, land use, and vegetation distribution in Libya. It also provides the results that are easily understood by the decision makers for better land use planning, improving the implementation of management strategies to meet the increasing demands for basic human needs and welfare of the ever-growing population in Libya.

6.2 Land Cover Map

The Libyan land cover dataset was created with a visual interpretation of the satellite imagery covering the period (2001–2002), using GeoVIS techniques and Land Cover Classification System (LCCS), followed by a dense field check operation using a series of field surveys all over the country. At the end of December 2005, the Libyan land cover database was updated with additional information gathered from precision assessment activities, all within the framework of the LIB/00/004 "Mapping of Natural Resources for Agriculture Use and Planning in Libya Project". Landsat satellite imagery of the 1980s, 1990s, and 2000 were compared using change detection analysis and followed by a dense field check operation. The final land cover map was created by clustering over a hundred of the original land cover classes into only ten generalized classes based on the most sensitive land cover features in Libya (Fig. 6.1 and Table 6.1). The result is an aggregation that keeps a good level of information, giving an accurate estimate of the areas covered by each aggregated class. Presently, this map and its accompanying statistics are one of the most reliable sources of the land cover in Libya, where it is also used by the provincial agencies, public, and private stakeholders for different purposes including development and conservation projects. The final aggregated land cove classes of this map can be summarized as follows (LIB/00/004 2009).

1. **Natural Forest and Reforest**: These forests originated from the original forest cover, which reproduced naturally. They have spontaneously generated themselves on the location and consist of natural immigrant tree species and strains.
2. **Irrigated Agriculture**: These areas cultivated the artificial application of water to the soil in order to grow crops. They are primarily used to replace missing rainfall in periods of drought, as well as to protect plants against frost.
3. **Rainfed Agriculture**: These areas include the cultivation of crops under conditions of natural rainfall.
4. **Rangeland**: This area contains a significant proportion of natural vegetation including native grasses, grass-like plants, and shrubs. Rangeland also consists of seeded areas to native or adapted introduced species that are managed like native vegetation.
5. **Sabkha**: These areas consist of flat depression and very saline areas of sand or silt lying close or above the water table and covered with a salt crust. Sabkha often contains soft nodules and enterolith veins of gypsum or anhydrite.
6. **Bare Soil in Wadi**: This area includes dry wadis with soil surface devoid of any plant material. Typically, it is either a drought environment or receives only a seasonal runoff in small quantities.
7. **Bare Soil Consolidated which known locally as Hamada (Rocky Deserts)**: These areas are characterized by the solid and firm consistency of their surface or by the presence of coarse fragments with these properties. The surface and the coarse materials remain coherent and hard even when moist.
8. **Bare Soil Unconsolidated which known locally as Sarrier (Desert Pavements)**: A defined area is covered with materials that are neither solid nor firm. The surface can be penetrated with a spade or a hoe.
9. **Loose and Shifting Sand**: These areas are covered by soil particles. These particles may be moved by regularly occurring winds and form distinct patterns, where a stony surface can be specified.
10. **Urban Areas**: These areas are nonlinear built-up areas covered by impervious structures adjacent to or connected by streets. This land cover is related to centers of population.

The use of Geographic Information Systems (GIS) technology helped to generate an accurate land cover map and illustrate statistically the relation of both the natural vegetation zones and the agricultural area within the desert ecosystem in Libya. As indicated in Table 5.1, the total area of Libyan land is about 1,759,540 km^2, where the

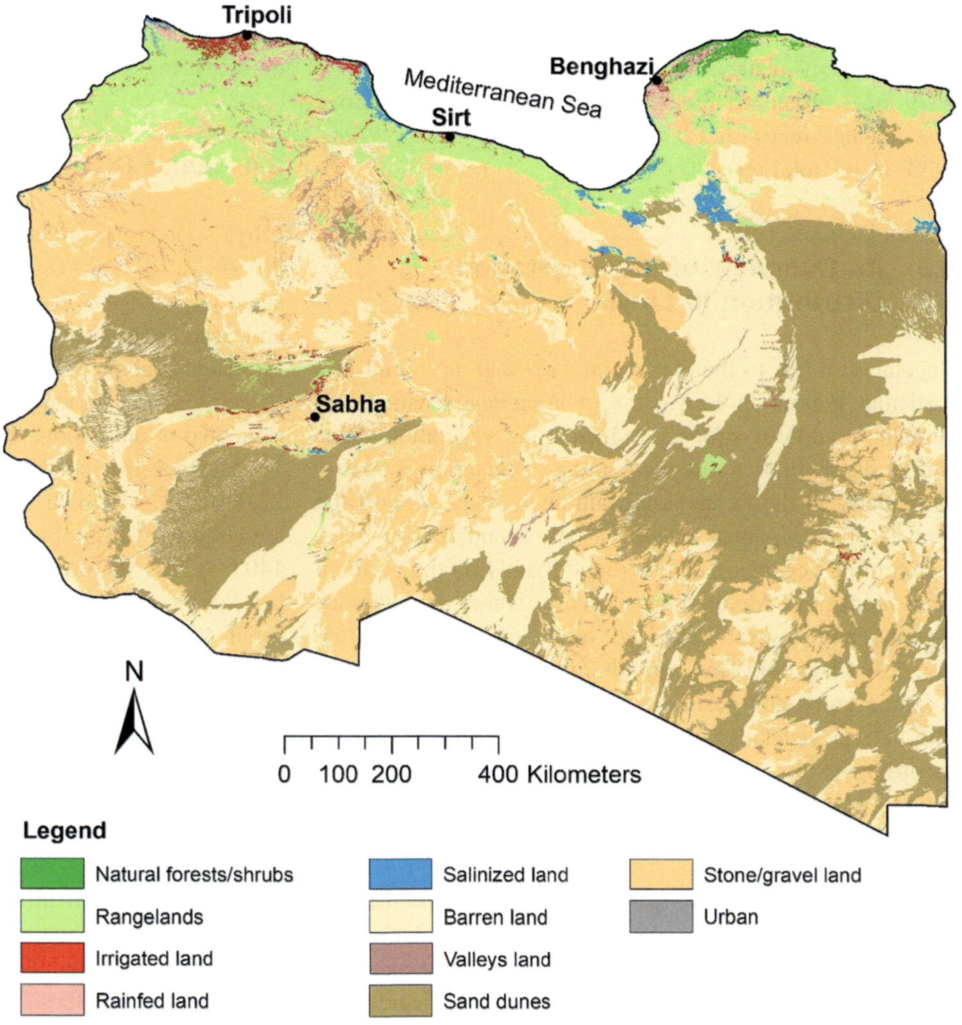

Fig. 6.1 The predominant land cover types in Libya. Adapted from LIB/00/004 (2009)

Table 6.1 The predominant land cover types in Libya based on (LIB/00/004 2009)

Land cover type	Total area (km^2)	Total area (%)
Irrigated land	6095	0.35
Rainfed land	14,873	0.85
Stone gravel land	680,623	38.68
Sand dunes	447,538	25.43
Barren land	429,961	24.44
Valleys land	15,471	0.88
Natural forests shrubs	3356	0.19
Rangelands	150,061	8.53
Salinized land	9679	0.55
Urban	1883	0.11
Overall total	1,759,540	100.00

agricultural area is 20,968 km^2 (irrigated 29%, and rainfed 71%), natural forests and shrubs are 3,356 km^2, rangelands is 150,061 km^2, valleys land is 15,471 km^2, salinized land is 9679 km^2, urbanized area is 1,883 km^2, and other stone gravel, sand dunes, and barren land around 1,558,122 km^2 (90% of the total area).

6.3 An Overview of the Vegetation Distribution in Libya

Vegetation refers to the plants and trees that are found in a particular region. In Libya, the geographical location and climatic conditions strongly influence the vegetation status and its distribution across the country. Vegetation is sparse, particularly in the desert areas, although oases support the growth of date palms, as well as olive and orange trees. Agricultural vegetation can be found mainly on the northern strip along the Mediterranean coast, some mountainous regions, and several dispersed oases in the desert (see Fig. 1.3, Chap. 1). This agricultural vegetation consists of common crops and pasture grasses as shown in Fig. (6.2). The natural vegetation distribution in Libya varies mainly according to the climatic zone and soil types. It includes natural forests of different trees and some other natural vegetation with predominant species of nonwoody shrubs as shown in Fig. 6.3.

The diversity of natural vegetation communities and native species of Libyan plants reflect local and regional variations in environmental factors. Acknowledging that most of the country is desert dry areas, the fertile regions are mainly present in the coastal strip areas. According to the Atlas of Libya (1978), the natural vegetation in Libya can be represented by twenty-one vegetation types (Fig. 6.4 and Table 6.2).

As shown in Fig. 6.4, the coastal region is characterized by halophytic vegetation, whereas the mountain areas prevail with Mediterranean vegetation, such as Aleppo pine forests, Arar forests with Junipers, and steppe with coastal forest belts. The subdesert and desert areas are characterized by the presence of steppe with bushes and shrubs and herbaceous plants in the subdesert areas, and sporadic vegetation and some short-lived vegetation after rains in the desert areas.

It was also stated in the Fourth National Report to the Convention on Biological Diversity that the biological diversity in Libya is considered relatively poor (according to the registered species) despite the large area of the country (Department of Environment, Tripoli, Libya 2010). Past studies and research have indicated that there are about 1750 plants in the country that follow 744 genera distributed over 118 families, of which only about 4% are endemic, such as the *Pachyctenium*, *Libyella*, and *Oudneya*. It is also worth mentioning that more than 50% of the total number of endemic plants can be found in Al Jabal Al Akhdar, in the northeast region of the country.

6.4 Land Distribution, Use, and Suitability

Libya's land area is about 1.76 million km^2 (Zurqani et al. 2019). More than 90% of this area is desert; only a very small percentage of Libyan land is capable of being plowed and used to grow crops. In 2016, the agricultural land is estimated at about 8.73% of the total area of the country with an arable land of about 1%. Land under permanent crops is estimated at only about 0.19% (Table 6.3).

In Libya, the suitable area for cultivation approximates 22,000 km^2, of which 2,390 km^2 are dedicated to irrigated agriculture, 15,500 km^2 to rainfed farming, and 140,000 km^2 to forest and rangelands. Irrigation potential has been estimated at about 7,500 km^2, relying mostly on fossil water, but when considering renewable water resources, it decreases to 400 km^2 in the coastal areas (Azzabi 1996). Generally, the crop yields of both rainfed and irrigation are low (Table 6.4). For example, the average yields of rainfed for wheat and barley are 650 kg/ha and 450 kg/ha, respectively. The average yields of irrigation for them are about 750 kg/ha and 1400 kg/ha, respectively.

The average yield of irrigated wheat and barley in Libya is much lower than the yields obtained in other Mediterranean countries. The yields of irrigated fruits and vegetable crops are generally also lower than in the surrounding countries, but for these crops, the differences are smaller. Besides the aridity of the climate, which reduces rainfed yields, this is due to prevailing shallow, coarse soils with limited natural fertility and high risk of soil erosions.

Determining how to use land efficiently in Libya is an ongoing need. Previous attempts to improve soil management in Libya have used a Productivity Rating Index (Ben Mahmoud 1995). This index is a parametric system that was adapted and developed for Libyan soils by Prof. Khaled Ben Mahmoud in 1995 (see Chap. 2). In 2013, it was integrated within Geographic Information Systems (GIS) to determine the productivity rating and suitability classes for selected soil uses in the northwestern of the country (Nwer et al. 2013). The results showed that olive and tomato are soil uses with the largest area of highly suitable class, whereas palm trees, potato, barley, and wheat are crops with the largest area with a marginally suitable class (Fig. 6.5 and Table 6.5).

Nwer et al. (2013) indicated that the use of a soil productivity rating Index that takes into account soil factors helps to distinguish the most limiting of these factors. Consequently, the results of soil suitability classes assisted in identifying which of the studied crops are suitable to grow and in which location. This evaluation can help decision

Fig. 6.2 Agriculture vegetated land; **a** Irrigated trees with herbaceous crop, **b** Herbaceous crops with pivot irrigation, **c** Irrigated herbaceous crops, **d** Irrigated crops with sparse shrubs, **e** Rainfed and irrigated crops with natural vegetation, and **f** Rainfed trees

Fig. 6.3 Natural vegetated land **a** Salt crust in wet sabkha environment, **b** Managed and-or natural palm trees, **c** Closed to open shrubland, **d** Very open woodland, **e** Sparse dwarf shrubs in pre-desertic area, and **f** Stoney bare soil with sparse vegetation

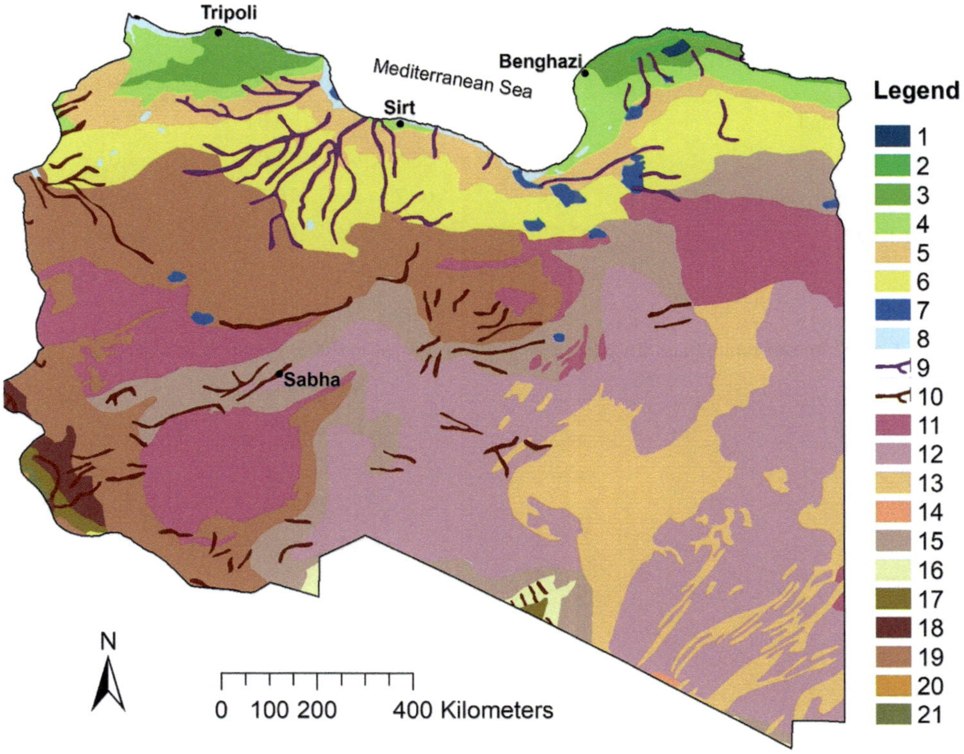

Fig. 6.4 The distribution of the natural vegetation types in Libya. Adapted from: Atlas of Libya (1978)

Table 6.2 The distribution of the natural vegetation types in Libya based on (Atlas of Libya 1978)

Color code	Natural vegetation type	Area (%)
1	Mediterranean evergreen woodlands	0.06
2	Open Aleppo pine forests	0.19
3	Open Arar forests with Junipers	1.93
4	Steppe with coastal forest belts	2.88
5	Sub-desert steppe with bushes and shrubs	4.39
6	Sub-desert steppe with herbaceous plants	8.97
7	Salt shotts	0.56
8	Salt marshes with shrubs	0.48
9	Mediterranean wadi vegetation with Acacias	1.58
10	Desert wadi vegetation with Nerium oleander	1.86
11	Desert dunes with some vegetation (Ephedra etc.)	12.70
12	Desert with almost no vegetation	25.96
13	Desert dunes without vegetation	10.26
14	Desert dunes with some vegetation (Aristidia etc.)	0.05
15	Short-lived, sporadic vegetation after rains	8.60
16	Tibisti lower mountain belt with Acacias	1.47
17	Tibisti hamada with Acacia flava	0.41
18	Ahaggar mountain desert steppe	0.40
19	Ahagger steppe with Acacias and Panicum	0.72
20	Ahaggar lower mountain belt with Acacias	16.46
21	Tibisti shrub grasslands	0.01

Table 6.3 Land use type based on (FAO 2020)

Land use type	Year						
	1961	1971	1981	1991	2001	2011	2016
Agricultural land	6.36	7.53	8.64	8.80	8.79	8.74	8.73
Arable land	0.97	0.98	1.00	1.03	1.03	0.98	0.98
Land under permanent crops	0.15	0.17	0.19	0.20	0.19	0.19	0.19
Other land	92.52	91.31	90.17	89.97	89.99	90.10	90.10

Table 6.4 Estimated rainfed and irrigated crop yields in Libya, in 2000, based on (FAO 2005)

Crop type	Yield in kg/ha	
	Rainfed	Under irrigation
Wheat	650	1,400
Barley	450	750
Dates	2,800	8,600
Potatoes	–	7,300
Pulses	600	1,500
Citrus	–	10,500
Apples	8,300	20,000
Grapes	2,300	10,400
Vegetables	6,700	13,000
Olives	700	2,200
Groundnuts	–	1,800

makers recognize the most limiting soil factors. Assessing the potential for improvement of manageable limiting soil factors such as soil salinity, exchangeable sodium, and drainage may increase the level of improvement in soil suitability.

6.5 Crop Production and Cropping Pattern

6.5.1 Crop Production

In general, cultivation in Libya is mostly constrained by very scarce freshwater resources, limited agricultural land, and low soil fertility. Consequently, it is characterized by an extensive production system with low productivity. Agricultural activities in Libya typically rely on groundwater to meet their irrigation needs (FAO 2015). Private farms owned by individuals provide the biggest part of the agriculture products in the country. Some other projects supported by the government were established under the irrigation system in the desert; these projects are predominantly for cereal and forage production (Al-Idrissi et al. 1996). Crop productions essentially depend on the private sector, where most of the farms are relatively small (5–20 ha), which limits the cultivated areas with grains (barley and wheat), vegetables (tomatoes, potatoes, onions, carrots, and turnips), fruits (olives, dates, peaches, amniotes, melons, grapes, and oranges), and forage crops. Figure 6.6 shows the values of the main produced crops in Libya (Zurqani et al. 2019).

Nearly 80% of Libya's agricultural production consists of fruit sand vegetables grown for domestic consumption. Climatic conditions limit grain production to barley and wheat. Due to the extremely low precipitation rate, most agriculture (over 80%) is heavily reliant on irrigation. Approximately 75% of Libya's food is imported, and agricultural activity is almost entirely for domestic consumption. In 2019, cereal production was estimated at 219,000 tons, which is almost exactly the same as in the previous year, but about 15% below the average. The country relies heavily on imports (up to 90%) to cover its grain consumption needs, mostly wheat and barley. In the marketing year 2020 (July/June), the actual import requirements are expected to reach 3.2 million tons, which is the same rate as the previous year (FAO 2019).

6.5.2 Cropping Pattern

The cropping pattern of Libya is mainly based on barley and wheat crops. They dominate the major cropping pattern

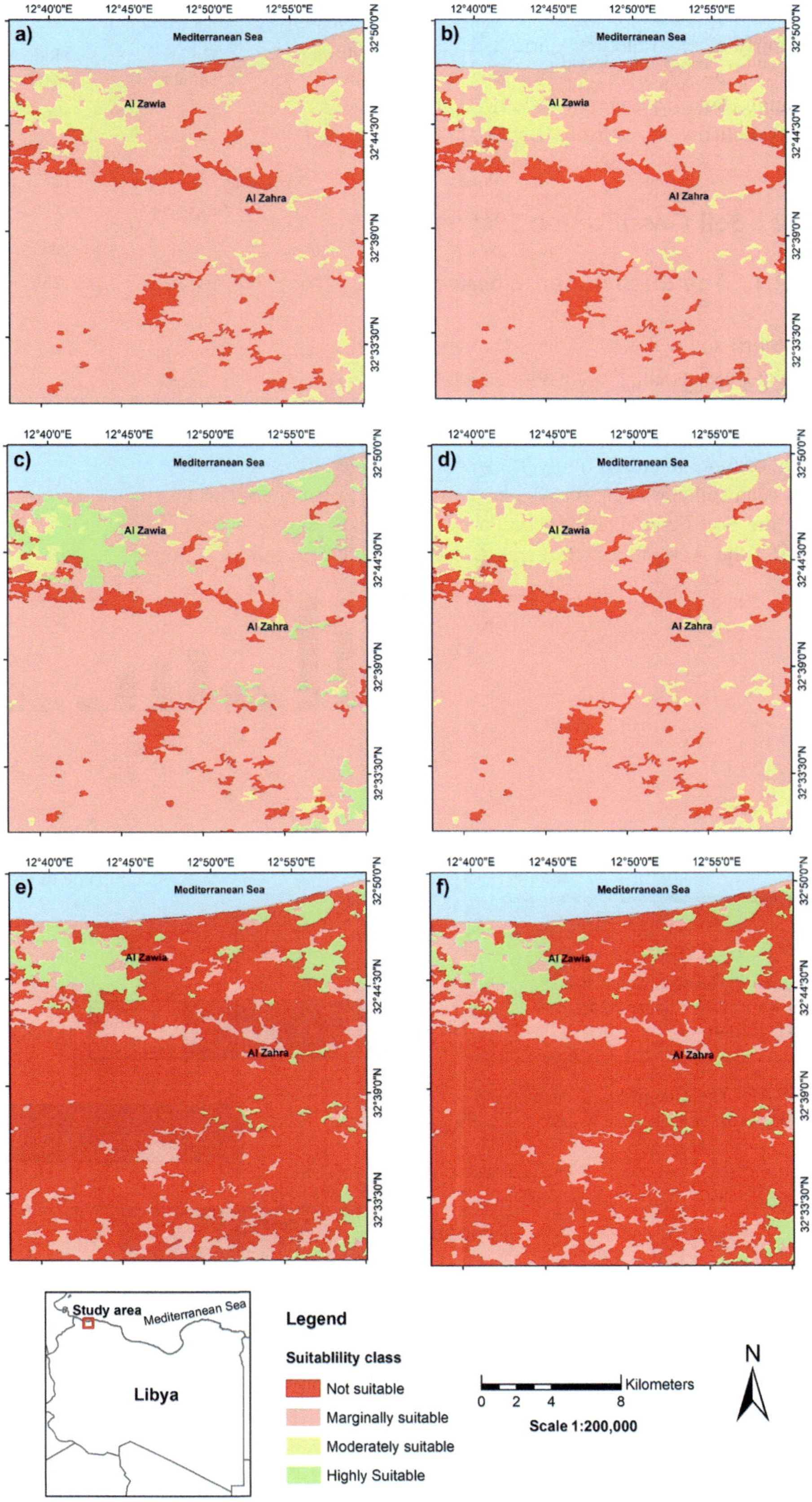

Fig. 6.5 Soil productivity rating index and suitability classes for selected crops in the northwestern, Libya; **a** Wheat, **b** Barley, **c** Tomato, **d** Potato, **e** Olive trees, **f** Palm trees. Adapted from: Nwer et al. (2013)

Table 6.5 The distribution of soil productivity ratings and suitability classes for selected crops in the northwestern, Libya based on (Nwer et al. 2013)

Crop type	Soil productivity ratings (Area ha)				
	Excellently suitable	Highly suitable	Moderately suitable	Marginally suitable	Not suitable
Barley	–	–	9658.75	97536.53	8326.3
Palm	–	–	9658.75	96559.22	9303.6
Potato	–	–	12320.55	94829.48	8371.55
Olive	–	9658.75	–	16742.08	89120.75
Wheat	–	–	9658.75	97536.53	8326.3
Tomato	–	9658.75	2661.8	95247.45	7953.58

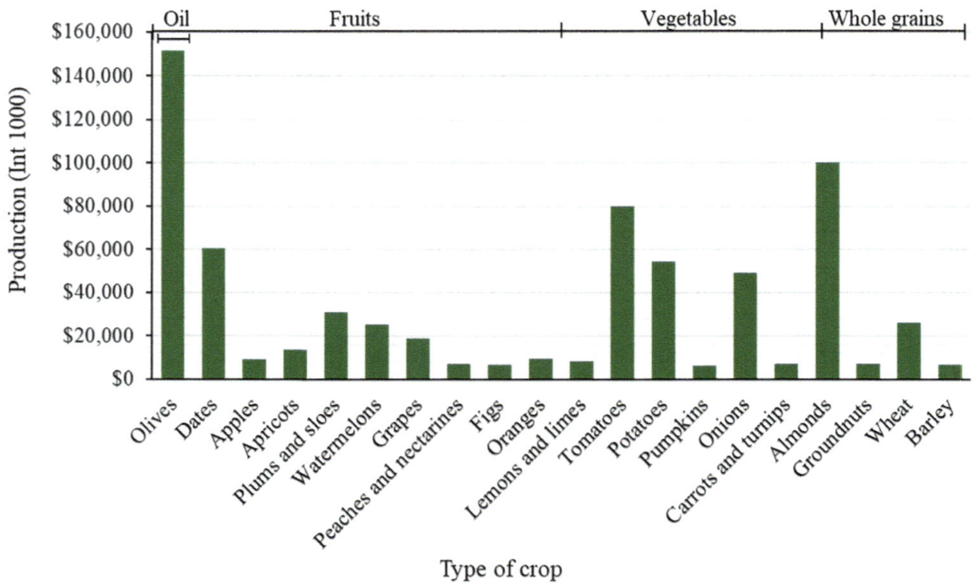

Fig. 6.6 Values of the main produced crops in Libya in 2016 based on (FAO 2020)

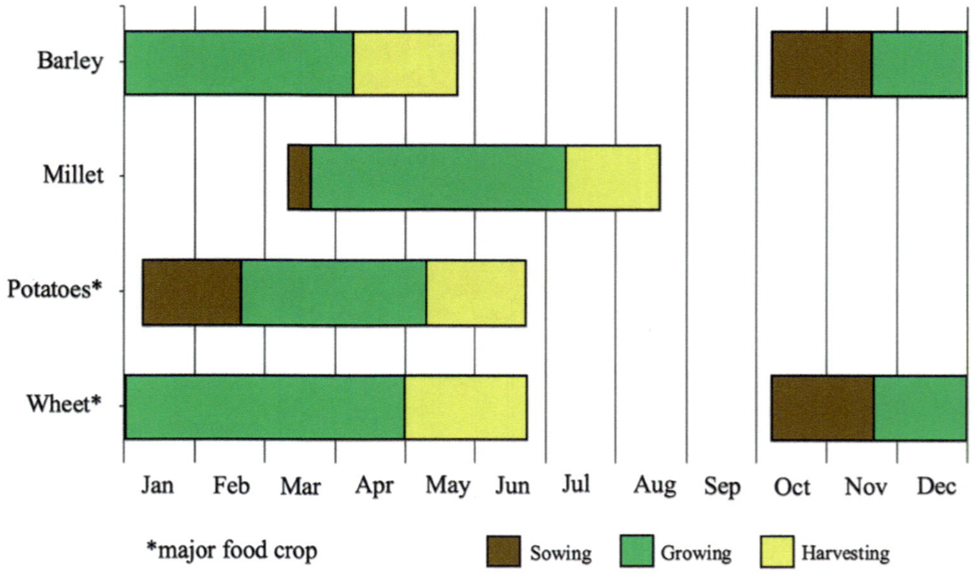

Fig. 6.7 The crop calendar is shown in Libya. Adapted from: FAO (2019)

throughout Libya. However, barley is considered as the main cereal crop across the country, as the lack of rain and low soil fertility give it an edge over wheat. Under favorable weather conditions, barley and wheat sowing begin in October. Harvesting of barley starts in April and ends in mid-May, while harvesting of wheat starts in May and is finalized by mid June (Fig. 6.7).

Available data from FAOSTAT indicate that the average harvested area graph is dominated by two cultivations, barley and wheat, while potatoes and millet account for only 4%, and 1%. Based on this data, wheat also displays the largest average production compared with the other crops (Fig. 6.8).

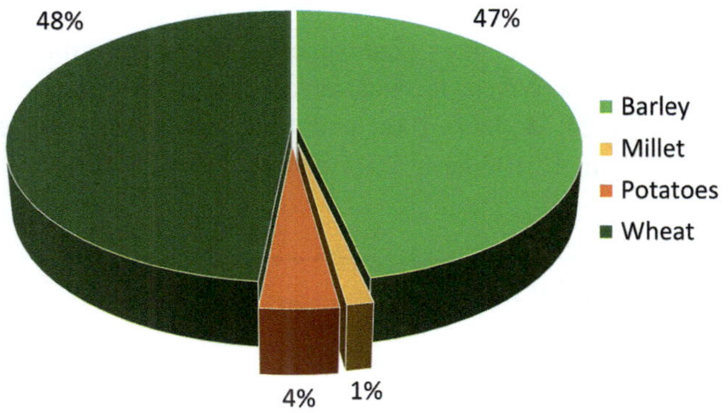

Fig. 6.8 Average area harvested (2000–2017) for the major crops in Libya based on (FAO 2020)

6.6 Forests and Shrublands

Libya is very lightly forested, with approximately 400,000 ha (0.2% land area) and some other wooded lands that cover about 446,000 ha (0.3%) (FAO 2004). There are no forests maintained for commercial purposes in the country. Generally, the natural forest mainly exists in the Al Jabal Al Akhdar, the northeast region, Benghazi; this region is characterized by forest tree species, such as juniper (*Juniperus spp.*) and mastic (*Pistacia spp.*). In the north-western part (Jabal Nafusa plateau), some shrublands can be found with predominant species of nonwoody shrubs, such as asphodel (*Asphodeline spp.*) and wild pistachio (FAO 2004). Table 6.6 shows a list of the scientific names for the dominant forest trees and shrubs in the country.

These forested and shrubland areas have occasionally suffered from severe deforestation, especially nowadays

Table 6.6 List of the scientific names for the dominant forest trees and shrubs in Libya based on (Ben Mahmoud 1995)

No	Scientific name	No	Scientific name
1	Juniperus Phoenicea	24	Zizyphus Lotus
2	Lygeum Spartum	25	Teucrium Polium
3	Hammada Scoparia	26	Calycotome Villosa
4	Pistacia Lentiscus	27	Globularia Alyoum
5	Artemisia Herbo alba	28	Thymelaea Hirsuta
6	Rhus Tripartita	29	Erica Multiflora
7	Retama Raetam	30	Lycium Europaeum
8	Quercus Coccifera	31	Tamarix Aphyliar
9	Stipagrostis Pungens	32	Morrubium Vulgaire
10	Arbutus Pavarii	33	Phlommis Floccosal
11	Artemisia Monosperma	34	Rosmarinus Officinalis
12	Cupressus Sempervirens	35	Salvia Trilobal
13	Anabasis Articulata	36	Thymus Capitatus
14	Pinus Halepensis	37	Nerium Oleander
15	Atriplex Mollis	38	Helianthemum Sp.
16	Ceratonia Siliqua	39	Sylphium Sp.
17	Olea Europaea	40	Pituranthos Torthosts
18	Arthrocnemum Glaucun	41	Orginea Martima
19	Suadea Fruticosa	42	Gymnocorpos Decider
20	Poterium Spinosum	43	Hammda Schmittian
21	Zygophyllun Album	44	Atractylis Cancellata
22	Ricinus Communis	45	Helianthenri Kahiricun
23	Cistus Parviflorus	46	Stipat Tenancissinal

where it is considered one of the biggest environmental challenges in the country (Zurqani et al. 2019). According to Global Forest Watch, the total area covered by these trees was about 24,344 ha in 2000, and it decreased over time, as it was about 12,107 ha in 2011 and 11,866 ha in 2018.

6.7 Conclusions

The harsh climate in arid and semiarid environments has a strong impact on vegetation and its distribution in these regions. Generally, arid and semiarid regions are unable to support forests or widely extended vegetation cover due to limited precipitation. In Libya, the land cover distribution and the vegetation types are highly correlated with geographical location and climatic zones. Most agricultural activities occur on a narrow strip along the Mediterranean coast, low mountains, and scattered oases in the desert. Crop productions are also quite variable across the country and typically depend on the private sector. In terms of natural vegetation, it can be represented by twenty-one vegetation types such as natural forests of different trees and some other natural vegetation with predominant species of nonwoody shrubs. Accordingly, quantifying and monitoring the spatial and temporal extent of the Libyan land cover is critical. Thus, updating the existing land cover map (LIB/00/004) more frequently is essential and may return further effective management strategies in the future for promoting sustainable land use planning.

References

Al-Idrissi M, Sbeita A, Jebriel A, Zintani A, Shreidi A, Ghawawi H (1996) Libya: Country Report to the FAO International Technical Conference on Plant Genetic Resources. Libya Country Report. Available online: www.fao.org/fileadmin/templates/agphome/documents/PGR/SoW1/east/LIBYA.pdf. Accessed 11 Dec 2019

Azzabi T (1996) Food self-sufficiency and agricultural research in Libya. Cahiers Options Mediterraneennes 1(5)

Ben Mahmoud KB (1995) Libyan soils, 1st ed. National Research Scientific Organization, Tripoli, Libya (Arabic book)

Chen D, Pan Q, Bai Y, Hu S, Huang J, Wang Q, Naeem S, Elser JJ, Wu J, Han X (2016) Effects of plant functional group loss on soil biota and net ecosystem exchange: a plant removal experiment in the Mongolian grassland. J Ecol 104(3):734–743

Department of Environment, Tripoli, Libya (2010) The fourth national report to the convention on biological diversity (Arabic Document)

Food and Agriculture Organization of the United Nations (FAO) (2004) Forests and the forestry sector. http://www.fao.org/forestry/country/57478/en/lby/. Accessed 12 Dec 2019

Food and Agriculture Organization of the United Nations (FAO) (2005) AQUASTAT Country Profile: Libyan Arab Jamahiriya. Land and Water Division. Food and Agriculture Organization of the United Nations, Rome, Italy. http://www.fao.org/nr/water/aquastat/countries_regions/lby/LBY-CP_eng.pdf. Accessed 18 Apr 2019

Food and Agriculture Organization of the United Nations (FAO) (2015) Status of the World's Soil Resources (SWSR)–Main Report. In: Food and agriculture organization of the united nations and intergovernmental technical panel on soils. FAO, Rome, Italy, p 650

Food and Agriculture Organization of the United Nations (FAO) (2019) GIEWS—Global information and early warning system, country briefs: Libya. Food and Agriculture Organization of the United Nations, Rome, Italy. http://www.fao.org/giews/countrybrief/country.jsp?code=LBY. Accessed 19 Apr 2020

Food and Agriculture Organization of the United Nations (FAO) (2020) FAOSTAT Statistical Database, Libya. http://www.fao.org/faostat/en/#data. Accessed 18 Apr 2020

Government of Libya, Secretariat of Planning, Surveying Department (1978) The National Atlas of Libya. Tripoli, Libya

LIB/00/004 (2009) Mapping of natural resources for agriculture use and planning in Libya Project, Atlas of natural resources for agricultural use in Libya. http://www.fao.org/geospatial/resources/detail/en/c/1024739/. Accessed 11 Apr 2019

Mahmoudzadeh A (2007) Vegetation cover plays the most important role in soil erosion control. Pak J Biol Sci 10(3):388–392

Maselli F, Di Gregorio A, Capecchi V, Breda F (2009) Enrichment of land-cover polygons with eco-climatic information derived from MODIS NDVI imagery. J Biogeogr 36(4):639–650

Nwer B, Zurqani H, Jadour K (2013) Soil productivity rating index model using geographic information system in Libya. In: Proceedings of the Annual International Conference 7th Edition of Geotunis, southern Hammamet, Tunis, 4–12 Apr 2013

Ruiz-Colmenero M, Bienes R, Eldridge DJ, Marques MJ (2013) Vegetation cover reduces erosion and enhances soil organic carbon in a vineyard in the central Spain. CATENA 104:153–160

The Statistical Office of the European Union (EUROSTAT) (2001) Manual of concepts on land cover and land use information systems. Office for Publications of the European Communities, Luxembourg, p 93. https://ec.europa.eu/eurostat/ramon/statmanuals/files/KS-34-00-407-__-I-EN.pdf. Accessed 18 Apr 2019

Zurqani HA, Mikhailova EA, Post CJ, Schlautman MA, Elhawej AR (2019) A review of Libyan soil databases for use within an ecosystem services framework. Land 8(5):82

Dr. Hamdi A. Zurqani is one of the faculty members at the Department of Soil and Water Sciences, Faculty of Agriculture, University of Tripoli, Tripoli, Libya. Dr. Zurqani is a recognized expert as a result of his internationally acclaimed work in the areas of environmental information science, remote sensing, land evaluation, sustainability, pedology, and soil science education. He has conducted research across the world, including the United States of America and Africa. Dr. Zurqani is a distinguished soil scientist with a wide range of scientific and working experiences in Libya and abroad. He received his M.Sc. (2010) from the University of Tripoli, Tripoli, Libya and Ph.D. (2019) from Clemson University, Clemson, SC, USA. His major research and teaching activities at the University of Tripoli have focused mainly on soil genesis and classification and the environmental information sciences (Remote Sensing and Geographic Information System). He has published broadly in many journals (e.g., Nature "Scientific Reports", Geoderma; International Journal of Applied Earth Observation and Geoinformation; Journal of Geological Sciences; Land; Frontiers in Environmental Science; Communications in Soil Science and Plant Analysis; and others). Dr. Zurqani is a member of the Editorial Board for Remote Sensing (MDPI) Journal, counseling outcome and research evaluation. He also was appointed to serve as a Guest Editor for the Special Issue "Applications of Remote Sensing in Earth Observation and Geo-Information Science". In addition, Dr. Zurqani conducted peer-review for many journals including Journal of Environmental Informatics, Applied Sciences, SN Applied Sciences, Remote Sensing, Heliyon, Geosciences, Land, Water, Agronomy, Agriculture, Sustainability, Arid Land Research and Management, International Journal of Environmental Research and Public Health, Natural Hazards, and Conference of the Arabian Journal of Geosciences. He is also one of the authors of the lab manual entitled, "GIS Exercises for Natural Resource Management". Dr. Zurqani has been the recipient of numerous awards and honors: Recipient of Douglas R. Phillips Award for Graduate Students, Department of Forestry and Environmental Conservation, Clemson University, April 12, 2019; the First Place Best Judged Poster (CAFLS) at the GRADS 2019: Clemson Student Research Forum on April 4, 2019; the Second Place Poster at the 11th Clemson Biological Sciences Annual Student Symposium, April 6, 2019; the Second Place Best Judged Poster at the Clemson Student Research Forum on April 4, 2018; and the Third Place Poster at the 9th Clemson Biological Sciences Annual Student Symposium, February 25, 2017. Dr. Zurqani conducts cutting-edge research in the field of environmental information science, remote sensing, land use management/planning, change detection of landscape degradation, and geographic information system (GIS) models. He has focused on his research efforts on the development of new technologies in the field of environmental information sciences, geo-intelligence (advanced geo-information science and earth observation, machine and deep learning, and big data analytics), remote sensing, land evaluation, pedology, land use management/planning, monitoring and evaluating sustainable land management, change detection of landscape degradation, and geographic information system models.

Dr. Khaled R. Ben Mahmoud is a senior professor of soil pedology. He is a distinguished soil scientist with wide range of scientific and working experiences in Libya and abroad. He received his Ph.D. in 1977 from Colorado State University (USA). His major teaching activities and research are at the Department of Soil and Water, Faculty of Agriculture, University of Tripoli, Libya. He has been teaching undergraduate and graduate students in Libya and has supervised many M.Sc. and Ph.D. thesis and dissertation in Libya and abroad. He focused mainly on soil resources, land evaluation, and mapping. He contributes to soil science in Libya and abroad as the Head of the Department of Soil and Water Sciences, Faculty of Agriculture, University of Tripoli, Tripoli, Libya (1978–1983), Dean of the Faculty of Agriculture, University of Tripoli, Tripoli, Libya (1983–1986), Chairman of the Committee for Libyan Agricultural Mapping, Tripoli, Libya (1988–1990). A Research Coordinator, University of Tripoli, Tripoli, Libya (1997–2001), Member of the Scientific Council of Agricultural Research Center (ARC), Tripoli, Libya (2000–2002), Member of the Scientific Council of the College of Higher Studies at the University of Tripoli, Tripoli, Libya (2000–2002), Member of the Scientific Committee of Libya Centre for Remote Sensing and Space Sciences, Tripoli, Libya (1998–2006), A National Coordinator of the Project implemented by FAO: Mapping of Natural Resources for Agricultural Use and Planning Tripoli, Libya (2000–2006), A General Coordinator of the National Programme for the Development of Vegetation Cover, Tripoli, Libya (2004–2006). Director of the Division of Land and Water Uses, The Arab Centre of the Studies of Arid Zones Lands (ACSAD), The League of Arab States, Damascus, Syria (2007–2011), Head of the Programme: Monitoring and Combating of Desertification in the Arab World, The Arab Centre of the Studies of Arid Zones Lands (ACSAD), The League of Arab States, Damascus, Syria (2007–2011), Director General of Eddayra Co. (Agricultural and Environmental Consulting and Services), Tripoli, Libya (2011–to date). Engaged as a local and international expert to control, evaluate, approve, and accept of many soils studies and mapping which have been implemented by different foreign and local companies, agencies, and offices in different parts of Libya and abroad (1977–to date). More than 100 country reports, research and scientific papers, and other publications were published in different local and international scientific journals and symposiums (in Arabic and English). He is the author of more than 10 books that have been written in Arabic. The most important ones are: Sandy soils of Libya (1981), The study of soils in the field (1984), Principles of Soil Science for Agricultural Secondary and Technical Schools (1985), Fundamentals of Soil Science and its Relation to Plant Growth (1995), Libyan Soils (Their Genesis, Classification, Properties, and Agricultural Potentiality (1995), Fundamentals of Soil and water Science (2001), Issues on Desertification and Land Degradation in the Arab Region (2010), Towards a National Strategy for the Sustainability of Natural Resources and Enhancing Food Security in Libya (2013), and The Libyan Desert (its natural potentials and prospects for investment) in press. Finally, he was appointed as a technical advisor to the President of the Libyan Presidential Council for Agriculture, Tripoli, Libya (2018–to date).

Soil Microbiology and Biotechnology

7

Eman Ali Ferjani, Merfat Taher Ben Mahmoud,
and Asma Yousef Alnajjar

Abstract

This chapter dealt with a brief presentation on the history of studying soil microbiology in Libya, the geographical location and climate, as well as the climatic factors that determine the vital activity in the soil. Then, research and studies that discussed the biological properties of soils and their environmental conditions were presented. Concerned with biological activity and modern techniques, to identify and classify microorganisms (bacteria and fungi) of most Libyan soils was one of the research priorities. Pathogenic fungi launched methods for chemical and biological control of pathogens, and research into pollution caused by pesticides and herbicides was also included. Truffles, wild pastures, and pastoral legumes, as well as their effect on the environment, have also been the focus of a share of the studies. Next, studies that focused on the positive aspects of soil microorganisms, such as the interest in symbiotic and non-symbiotic nitrogen fixation, were presented. This information was followed by a presentation of specific studies emphasizing ways to solve the most harmful environmental problems that represent the most significant challenges of biological activity (salinity, drought, heat, and pollution). Finally, the chapter provides research concerned with the bioremediation of oil residues and heavy elements and finds sound ways to dispose them of.

Keywords

Biodiversity • Microflora • Nitrogen fixation • Pesticides • Salinity • Bioremediation

7.1 Chronicle of Agricultural Education in Libya

In Libya, like other countries where interest in agriculture has appeared since ancient times, agriculture is the main source of food, despite the limited agricultural areas and difficult natural conditions. However, recent and rapid developments in society have made it necessary to pay more attention to the agricultural side in order to keep up with agricultural development plans and to cover the need for agricultural engineers, specialized technicians, and the preparation of highly specialized cadres. Agricultural education in Libya has gone through several stages throughout history and has transformed from the traditional farming system, in which the farmer instinctively learns and experiences, and in which agricultural information and foundations are passed down from one generation to another, to an educational and systematic agricultural system based on more modern foundations. Here, we will try to shed light on some important aspects of this change.

In 1909, when Libya was under Ottoman rule, interest appeared in agriculture as an institution for the first time in order to increase agricultural and animal production by commissioning and appointing the first commissioner of agriculture in Tripoli. A decision was made by the state council to establish a model farm in Mansheya, in the state of Tripoli, where a section for horticulture and another for animal production were allocated. This project included agricultural development, and experts were appointed in the agricultural education of farmers. After this model farm was created, the Al Awailia Institute for Agricultural Education was established in Al Marj in Al Jabal Al Akhdar (English:

E. A. Ferjani (✉) · M. T. Ben Mahmoud · A. Y. Alnajjar
Faculty of Agriculture, Department of Soil and Water Sciences, University of Tripoli, Tripoli, Libya
e-mail: eman.ferjani@gmail.com

M. T. Ben Mahmoud
e-mail: dr.mbenmahmoud@yahoo.com

A. Y. Alnajjar
e-mail: Aynsw2912@gmail.com

H. A. Zurqani (ed.), *The Soils of Libya*, World Soils Book Series,
https://doi.org/10.1007/978-3-030-66368-1_7

The Green Mountain), eastern Libya, as well as the Alghiran Agricultural Institute west of Tripoli (Alzobaidy and Mohammad 2003).

The Faculty of Agriculture at the University of Tripoli was established (June 1966) as one of the colleges of the Libyan University, and the study started at that time with three departments: the Soil and Water Department, the Department of Plant Production, and the Department of Animal Production. The first class, which consisted of a batch of 42 students, was accepted at the beginning of the 1966/1967 academic year. From this date, the start of the study of soil microbiology was one of the specializations and courses taught in the Soil and Water Department. After the first batch of 29 graduates in the year 1970, the number of graduates developed until the school's cumulative number in the period 2009/2010 reached 8771 graduates.

Beyond the University of Tripoli, other agricultural colleges were established in the succeeding years. In 1975, the Faculty of Agriculture was established at the University of Omar Al-Mukhtar in the Al Jabal Al Akhdar region in eastern Libya. The Soil and Water Department was one of the departments in the college. The Faculty of Agriculture was established in Sabha University in southern Libya in 1983, which numbered 815 students until 1996. In 1988, the College of Veterinary and Agricultural Sciences was established in Zawia, west of Tripoli. The Faculty of Agriculture at Misrata University was established and opened in 2015 in the city of Misrata, east of Tripoli. All of these colleges are studying soil microbiology as one of the courses in the Soil and Water Department.

7.2 Climate Diversity in Libya and Its Impact on Microbial Diversity

Libya occupies a wide area of 1,777,500 km^2 located between the latitude 32° 58′ and 18° 45′ N, and between the longitudes 9° 18′ and 24° 59′ E, which makes it characterized by a diverse climate. North of Libya is characterized by the climate of the Mediterranean Sea, which is represented in the coastal strip and extends a distance of 1900 km along the coast of the Mediterranean Sea. This area does not exceed 10% of the total area and differs in breadth from one place to another. The average rainfall in this region ranges from 300 to 600 mm, and average temperatures range between 25 and 40 °C in the summer and 0 and 14 °C in the winter, which allows vegetation to grow with different intensity and species according to the prevailing environmental conditions. As for most of the area located south of the coastal strip, which is estimated at about 90% of the total area, it falls within the desert climate or within the arid and semi-arid areas, which are characterized by the lack or scarcity of rainfall, and thus results in scarcity or lack of vegetation (Zekry 2005; and Environment Public Authority 2008). Although the Libyan soil, in particular, varies, in general it is very similar in its properties to many other dry areas.

The climate diversity, in turn, leads to biological diversity especially in the vegetation cover; therefore, microbial diversity exists, in particular within the distribution of microorganisms in soil layers in terms of quantity and quality, each according to the prevailing climatic conditions. The absence of vegetation and the scarcity of rain resulted in little or no organic matter in most arid areas, and since most of the microorganisms in the soil are heterotrophic species and depend entirely on organic matter, the microbial activity is almost non-existent in desert soils, which fall within the scope of arid and semi-arid soils. As for the areas that fall within the Mediterranean climate, the microbial activity depends quantitatively and qualitatively on the type of vegetation, the moisture content, and the percentage of organic matter, which are completely different from the dry areas.

7.3 Soil Bacteria in Some Regions of Libya

Microflora of rhizospheric soil and rhizoplane portions of *Aristida coerulescens* (naturally occurring in the Libyan desert) were different in counts and isolates in the different root zones. The root base contained the lowest numbers of microflora (bacteria and fungi), while the root tip included the highest counts (Naim 1965). Other investigations dealing with the effect of plants on the microbial rhizosphere population were extended to reveal the microbiological occurrence in different horizons of a soil profile corresponding in length of root regions. Bacteria were most abundant, followed by Actinomycetes, while fungi were less abundant and more restricted in their distribution. The three groups of microorganisms varied markedly in the rhizosphere of the plants. The plant's age had a great influence on the frequency of occurrence of the different microorganisms. One strain of bacteria, six fungi, and the members of the grey series of Streptomyces were dominant. Actinomycetes and fungi diminished with the depth of soil, while the bacterial counts increased (Selim and Khalil 1979). The results of these studies, which are considered somewhat old, agreed with a more recent study, which attempted to identify the effect of the growth of different plants on the microbial community in the soil of rhizospheres by tracking the numbers of three microbial groups (bacteria, fungi, and Actinomycetes) during the different growth stages (germination, flowering, and maturity), with two types of plant roots (leguminous and grass). It turns out that the numbers of bacteria and fungi in the rhizosphere of both plants increase as the plant ages, and then starts to decrease during the ripening stage. As for Actinomycetes, their numbers in the root zone increased in the last stages of growth (Al Hammadi 2014).

On the other hand, various strains of *Azospirillum* sp. were isolated from the rhizospheric soils of 23 leguminous and non-leguminous plants distributed in a unique Mediterranean-type climate in eastern Libya. Soils with the highest counts were observed in the rhizosphere of legumes. Based on the morphology of the cells and cultural and biochemical characteristics, 15 strains of *Azospirillum lipoferum* were identified. The distributions and characterizations of these strains varied based on the plant species and soil properties (Attitalla and Salleh 2010). *Azospirillum* was present in abundant in high population levels that recorded 1.1×10^3–1.32×10^5 CFU g^{-1} in 23 soil samples collected from the rhizosphere of 24 plants grown in different sites distributed in the Al Jabal Al Akhdar region (Attitalla 2011). *Azospirillum* was also found abundant in the rhizosphere of halophytes, which are collected from El-Thaliga salt marsh near Benghazi. More than 25 *Azospirillum* isolates were recovered both from free and rhizospheric soil samples. A total of 13 isolates were screened to select the most efficient strains in a microcosms experiment with wheat. Morphological and biochemical tests indicated that the three *Azospirillum* isolates belonged to *A. brasilense*, and two belonged to *A. lipoferum* (Taher et al. 2019). Different genera of cellulose-decomposing bacteria were isolated from the mentioned soils, including *Cytophaga*. Additionally, *Staphylococcus* was detected in most studied samples (Baiu and Nair 1999). While searching for antagonistic Actinomycetes in Libyan soils, 135 *Streptomyces* isolates from 39 soil specimens collected from different geographical regions of Libya have been determined. The isolated *Streptomyces* fall into four groups; the members of the grey series predominate, followed by the brown ones. The other series occur in negligible numbers. Sixty-five percent of the isolates proved to be antagonists, including 50% active exclusively against Gram-positive, 3% active exclusively against Gram-negative, and 12% active equally against Gram-positive and Gram-negative microorganisms (Selim 1971).

7.4 Surveying and Screening of Soil Fugal Flora in Some Regions of Libya

Mycoflora of soils have received considerable attention during the last 40 years; most of the work in this respect has been concerned with those fungi inhabiting cultivated, desert, and saline soils. Some investigations have been made into soil fungi in Libya. Naim (1976a, b) studied rhizosphere and soil fungi of *Artemisia herba-alba* and fungus flora under citrus trees in Tripoli, Libya. Youssef (1974) studied the fungal flora of Libyan soil. El-Said and Saleem (2008) studied soil fungi in the western region of Libya. Mansour (2010) studied soil fungi in the eastern region. Preliminary studies on isolating and defining fungi species of Libyan soils focused on rhizosphere and soil counts of fungi around *Artemisia herba-alba* grown in three localities in Tripoli. Counts from localities near the sea shore were higher than the counts from areas distant from the south. From this area, 10 genera and 25 species commonly presented in the rhizosphere or soil of *Artemisia* could be identified (Naim 1967). In another study, 63 species in 21 genera were reported. Four of these species were Phycomycetes, 10 were Ascomycetes, and 49 were Deuteromycetes. Of the reported species 55 were new records, and one of these was a new species (Youssef 1974).

Surveys and screenings of soil fungal flora have been made in wide areas of western and eastern regions of Libya, in addition to other considerable investigations in different parts of the country. Ecological and physiological studies on soil fungi in the western region of Libya, which covered a large area of the Jafara Plain, managed to isolate, estimate, and screen many genera and species of fungi (El-Said and Saleem 2008). A total of 75 soil samples of cultivated, desert, and saline soils (25 for each) were collected from different localities in the western region. The geographical feature of the Jafara Plain (the region of study) is referred to as a large area, as it is covers more than 17,000 km^2. It takes a triangle shape with the apex at the east, near Al khums town. The north is parallel with the Mediterranean Sea coast and is about 275 km long. The western side forms the western borders of Libya and is about 150 km long (see Fig. 1.1, Chap. 1). The highest value of total soluble salts was detected in saline soil (1.02–8.89%), in the cultivated (0.06–0.85%), and desert (0.01–0.48%) soils.

A total of 63 species and 5 varieties belonging to 30 fungal genera were recovered from the 75 different soil samples: cultivated (29 genera (gen.) and 58 species (sp.) + 5 varieties (var.)), desert (22 gen. and 35 sp. + 2 var.) and saline soil (21 gen. and 41 sp. + 1 var.). The most common genera were showed in Table 7.1.

Studying the distribution and occurrence of various groups of fungi in cultivated, desert, and saline soils in eastern region contributed to knowledge of more soil fungi (Mansour 2010). In this investigation, 100 soil samples were collected from different sites spread over 10 locations. The geographic area of the study varied over a wide area: from Ajdabiya at the south, to Benghazi and along the coastline of the Mediterranean Sea and the Al Jabal Al Akhdar to Derna at the east (see Fig. 1.1, Chap. 1). Water content varied between high dry soils (0.1–0.3%) and high wet soils (15–20.5%) and occurred in the rainy locations in Al Jabal Al Akhdar sites. The value of the total soluble salts was highest in saline soils (1–8.9%); these values were not found in the cultivated and desert soils. Fifty-nine fungal species belonging to 23 genera were obtained from 100 collected samples. The most common genera and species is shown in Table 7.2.

Table 7.1 Average total count (calculated per gram dry soil in every sample) for fungal genera and species recovered from soil samples cultivated on glucose, cellulose, and 50% sucrose-Czapek's agar at 28 °C (adapted from: El-Said and Saleem 2008)

No.	Genera and species	Average total count (calculated/g dry soil in every sample)		
		Cultivated	Desert	Saline
1	*Acremonium strictum*	100	40	120–1160
2	*Alternaria alternata*	100–340	300–2780	860–1900
3	***Aspergillus***	11880–32000	8780–21860	10340–22340
4	*A. candidus*	180–260	–	80–140
5	*A. carneus*	220	–	–
6	*A. flavus*	3480–9720	4880–10180	10300–8580
7	*A. flavus var. columnaris*	60–100	–	–
8	*A. fumigatus*	100–2280	960–4380	20–380
9	*A. niger*	4220–12380	340–6660	3700–10340
10	*A. ochraceus*	600–1180	220–260	280–860
11	*A. sydowii*	240–260	140	120–200
12	*A. terreus*	1200–6940	500–2180	1120–2220
13	*A. terreus var. africanus*	200–380	–	–
14	*A. terreus var. aureus*	540–600	120–940	680
15	*A. ustus*	920–2400	540–900	540–1200
16	*A. versicolor*	240	–	–
17	*A. wentii*	1300–2480	–	–
18	***Botryotrichum atrogriseum***	100–1540	–	80–360
19	***Chaetomium globosum***	140–740	880	100–1380
20	***Cladosporium***	140–400	–	–
21	*C. cladosporioides*	140–240	–	240–380
22	*C. sphaerospermum*	160	–	–
23	***Cochliobolus***	200–620	–	–
24	*C. hawaiiensis*	300	–	–
25	*C. lunatus*	100	–	–
26	*C. spicifer*	100–320	60	180–280
27	***Cunninghamella echinulata***	160		
28	***Emericella***	3560–11800	1120–4220	–
29	*E. nidulans*	3560–11100	1120–4220	1260–3500
30	*E. nidulans var. dentata*	460	–	–
31	*E. nidulans var. lata*	240	520	–
32	***Fusarium***	920–3620	–	680–1500
33	*F. dimerum*	820	–	280
34	*F. moniliforme*	300	–	180 - 620
35	*F. oxysporum*	700–3120	160	500–1220
36	*F. poae*	220–420	–	–
37	***Gibberella***	120–500	–	–
38	*G. acuminata*	120	–	–
39	*G. intricans*	500	–	–
40	***Humicola***	600–5760	–	–
41	*H. brevis*	160–1460	–	–
42	*H. grisea*	440–4300	480	–
43	***Eurotium***	15060	–	–

(continued)

Table 7.1 (continued)

No.	Genera and species	Average total count (calculated/g dry soil in every sample)		
		Cultivated	Desert	Saline
44	*E. amstelodami*	4540	–	–
45	*E. chevalieri*	10100	–	–
46	*E. rubrum*	420	–	–
47	***Mucor***	100–160	740–3320	540–760
48	*M. circinelloides*	60	640	–
49	*M. racemosus*	60–160	740–3320	220–620
50	***Mycosphaerella***	280–1460	80	2140–2760
51	***Myrothecium***	180	–	–
52	***Nectria haematococca***	10640–14080	120	900–7420
53	***Paecilomyces variotii***	260	–	–
54	***Penicillium***	2840–13640	7480–11280	7780–10960
55	*P. chrysogenum*	3980–12400	7120–11000	7100–10500
56	*P. citrinum*	160–360	140	60
57	*P. corylophilum*	220–300	–	100–460
58	*P. funiculosum*	240	–	–
59	*P. puberulum Bainier*	480–1240	80–440	280–580
60	*P. purpurogenum*	140	–	-
61	***Phoma glomerata***	100	60–260	40–200
62	***Pleospora herbarum***	100–540	240–720	100–780
63	***Rhizopus stolonifer***	80	120–200	60–80
64	***Scopulariopsis breviculis***	100	–	–
65	***Scytalidium lignicola***	100–180	80–300	780–2840
66	***Setosphaeria rostrata***	100–300	–	–
67	***Stachybotrys chartarum***	180–400	360	160–6460
68	***Syncephalastrum racemosum***	160–220	680–1060	–
69	***Torula herbarum***	60	520–620	100–1380
70	***Ulocladium***	100–880	400–720	80–180
71	***U. alternariae***	100–260	240–400	–
72	***U. botrytis***	240	420–500	80–100
73	***U. chartarum***	320	220	20–80
74	***U. tuberculatum***	60	–	80
	Gross total count	45940–75860	28820–61600	42960–58340
	Number of genera	29	22	21
	Number of species	58 + 5 var.	35 + 2 var.	41 + 1 var.

It is clear that there are great similarities in the genera of fungi and their different species isolated from the western and eastern regions. The studies revealed that the most prevalent species in both regions are: *Alternaria alternata, Aspergillus flavus, A. fumigatus, A. niger, A. terreus, Emericella nidulance, Fusarium oxysporium, Mycosphaerella tassiana, Nectria haematococca*, and *Penicillium chrysogenum.*

No rigid definition of soil fungi is given, but fungi detected or isolated from soil, seeds, and roots of plants are tentatively termed as soil fungi. Isolates from plant residues in soil, soil-borne animals, or mushrooms are also included as soil fungi (Watanabe et al. 1988 and Watanabe 2002). It is natural for soil fungi to be carried on the seeds of many plants; this helps in their transmission from one soil to

Table 7.2 Total fungal genera and species obtained from 100 soil samples collected from 10 localities in eastern region of Libya (adapted from: Mansour 2010)

No.	Genera	Species
1	*Alternaria*	*Alternaria alternata*
		Alternaria citri
		Alternaria humicola
		Alternaria tenuis
2	*Aspergillus*	*Aspergillus candidus*
		Aspergillus flavipes
		Aspergillus flavus
		Aspergillus fumigatus
		Aspergillus nidulans
		Aspergillus niger
		Aspergillus ochraceus
		Aspergillus sydowii
		Aspergillus terreus
		Aspergillus ustus
		Aspergillus sp
3	*Chaetomium*	*Chaetomium globosum*
		Chaetomium sp.
4	*Cladosporium*	*Cladosporium herbarum*
		Cladosporium cladosporioides
		Cladosporium herbarum
		Cladosporium macrocarpum
		Cladosporium sphaerospermum
5	*Cunninghamella*	*Cunninghamella echinulate*
		Cunninghamella sp.
6	*Eurotium*	*Eurotium* sp.
7	*Fusarium*	*Fusarium equiseti*
		Fusarium moniliforme
		Fusarium oxysporium
		Fusarium semitectum
		Fusarium solani
8	*Helminthosporium*	*Helminthosporium sativum*
9	*Humicola*	*Humicola* sp.
10	*Mortierella*	*Mortierella* sp.
11	*Mucor*	*Mucor hiemalis*
		Mucor racemosus
12	*Mycosphaerella*	*Mycosphaerella* sp.
13	*Myrothecium*	*Myrothecium roridum*
14	*Nectria*	*Nectria haematococca*
15	*Penicillium*	*Penicillium canescens*
		Penicillium chrysogenum
		Penicillium citrinum
		Penicillium cyclopium
		Penicillium decumbens
		Penicillium digitatum

(continued)

Table 7.2 (continued)

No.	Genera	Species
		Penicillium frequentans
		Penicillium funiculosum
		Penicillium lilacinum
		Penicillium thomii
16	*Phoma*	*Phoma exigua*
		Phoma glomerata
		Phoma sp.
17	*Phytophthora*	*Phytophthora* sp.
18	*Pleospora*	*Pleospora herbarum*
19	*Rhizoctonia*	*Rhizoctonia solani*
20	*Rhizopus*	*Rhizopus stolonifer*
21	*Trichoderma*	*Trichoderma viride*
22	*Trimmatostroma*	*Trimmatostroma* sp.
23	*Ulocladium*	*Ulocladium atrum*
		Ulocladium botrytis
	Total number of genera 23	**Total number of species** 60

another. This is especially true of the pathogens, which increases the transmission of infection. From this standpoint, several studies have gone into revealing the extent of infection of seeds with pathogenic fungi and have searched for ways to treat fungal infections with chemical and biological control.

Local leguminous seeds collected from different sites of Al Jabal Al Akhdar were tested for infection by pathogenic fungi accompanied with seeds. Fungal isolation from seed specimens represent infection of all tested seeds, with 18 different fungal species illustrated in Table 7.3.

Another study was carried out as an attempt for isolation and testing the pathogenicity of pathogenic fungi carried on some chickpea cultivars growing at Al Jabal Al Akhdar (Abdallah 2006). Twelve species of fungi (*Aspergillus ustus, A. candidus, A. niger, Cladosporium* sp., *Chaetomium bostrychodes, Fusarium oxysporum, Phomopsis* sp., *Penicillium frequentans, Pythium ultimum, Thielaviopsis* sp., *Rhizopus* sp., and *Rhizoctonia* sp.) were isolated from seeds samples. Pathogenicity was tested for isolated fungi on their hosts using a percentage of pre- and post-emergence damping off disease index (Fig. 7.1). The highest percentage of pathogenicity was recorded for *Fusarium oxysporum* and *Penicillium frequentans*.

Isolation and identification of fungi associated with food grains were carried out on 20 food grains including cereals (wheat, barley, corn, sorghum, fenugreek, millet); legumes (pea, peanut, chickpea, faba bean, kidney bean, cowpea, red and green lentil); oil-producing seeds (walnut, almond, hazelnut, sesame, pistachio); and two animal feeds (diet and ration) obtained from local markets during 2003–2006 in Al Bayda, Libya. Ten fungal genera were isolated; 20 species and two strains of *Saccharomyces cerevisiae* were isolated and identified. These species were *Aspergillus flavus, A. ochraceus, A. terreus, A. niger, A. candidus, A. clavatus, A. fumigatus, Penicillium chrysogenum, P. canescens, P. waksmanii, Fusarium oxysporum, F. graminearum, F. sporotrichioides, Rhizopus stolonifer, Mucor piriformis, Alternaria tenuissima, Rhizoctonia solani, Pythium ultimum, Phyllactinia rigida,* and *Saccharomyces cerevisiae*. The percentage of some fungal contamination is illustrated in Fig. 7.2.

The Bayoud disease caused by *Fusarium oxysporum, f.* sp. *albedinis* of the date palm tree is one of the most dangerous diseases in the world and is difficult to control. It has now spread in some countries of North Africa. In the frame of a regional project on Bayoud disease of date palm in 15 Arab countries, results showed that nearly all Arab soils present high to middle levels of receptivity to the fungus and that some soils are important, for example soil of Al Ghamr in Libya. This research gives an idea, not about the disease spread, but it permits us to imagine the map of spread risk of the disease according to counties that are still free and threatened by the contamination (Sedra 2010).

Chemical control of *Fusarium solani* isolated from Local Bean was treated by using five fungicides: *Roveral, Rizolex, Vitavax, Benlate,* and *Cabtan*. Results (Table 7.4) indicated that fungal growth was completely inhibited after 192 hours

Table 7.3 Different fungal species isolated from different leguminous seeds collected from EL-Jable Al Akhdar (adapted from: Abdallah et al. 2005a)

Fungi type	Legume type							
	Local Bean	Local Peas I	Local Peas II	Peas LS	Peas MG	Marj Lentil	Local Bean	Local Peas I
Alternaria alternata	+	–	–	–	–	–	+	+
A. tenuissima	+	+	–	–	–	–	–	–
Aspergillus ustus	–	+	–	–	–	–	–	–
A. flavus	+	–	–	+	–	–	–	–
A. niger	+	+	+	+	+	+	+	+
A. parasiticus	–	–	–	–	–	+	–	–
Fusarium oxysporum	+	+	+	–	–	+	–	+
F. solani	–	–	–	–	–	–	–	+
Nectria sp.	–	–	+	–	–	–	–	–
Phoma medicaginis	+	+	–	–	–	–	–	–
Penicillium fellutanum	–	–	–	+	–	–	–	–
P. janthinellum	–	–	–	–	+	–	–	–
Phytophthora sp.	+	–	–	–	–	–	–	–
Rhizopus sp.	+	–	–	+	+	+	+	+
Rhizoctonia solani	+	–	–	–	–	–	–	–
Stemphylium botryosum	–	+	+	–	–	–	–	–
Trichoderma viride	–	–	+	–	–	–	–	–

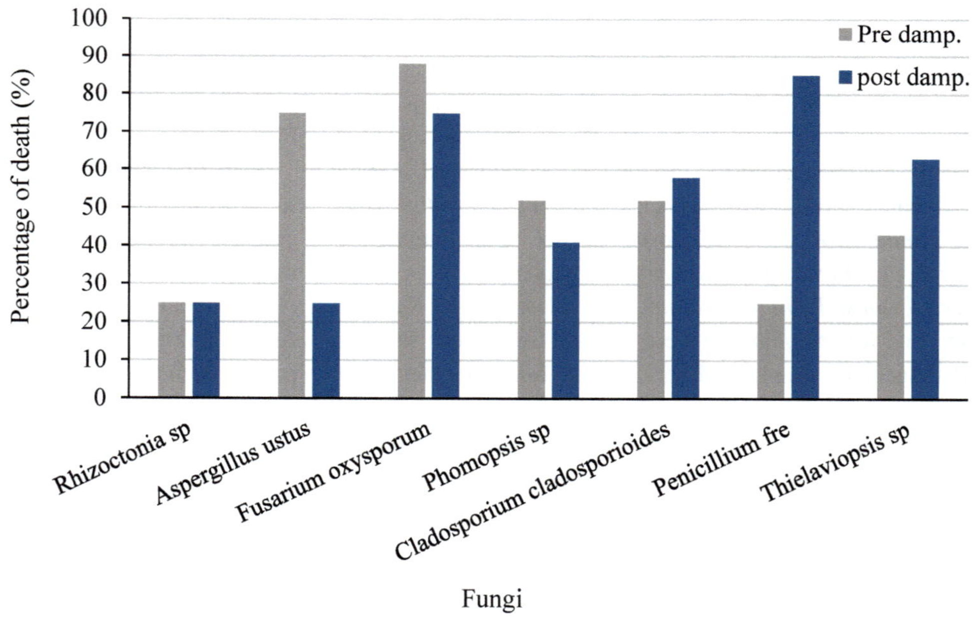

Fig. 7.1 The influence of isolated fungi on the percentage emergence death of seeds pre- and post-emergence damping (adapted from: Abdallah 2006)

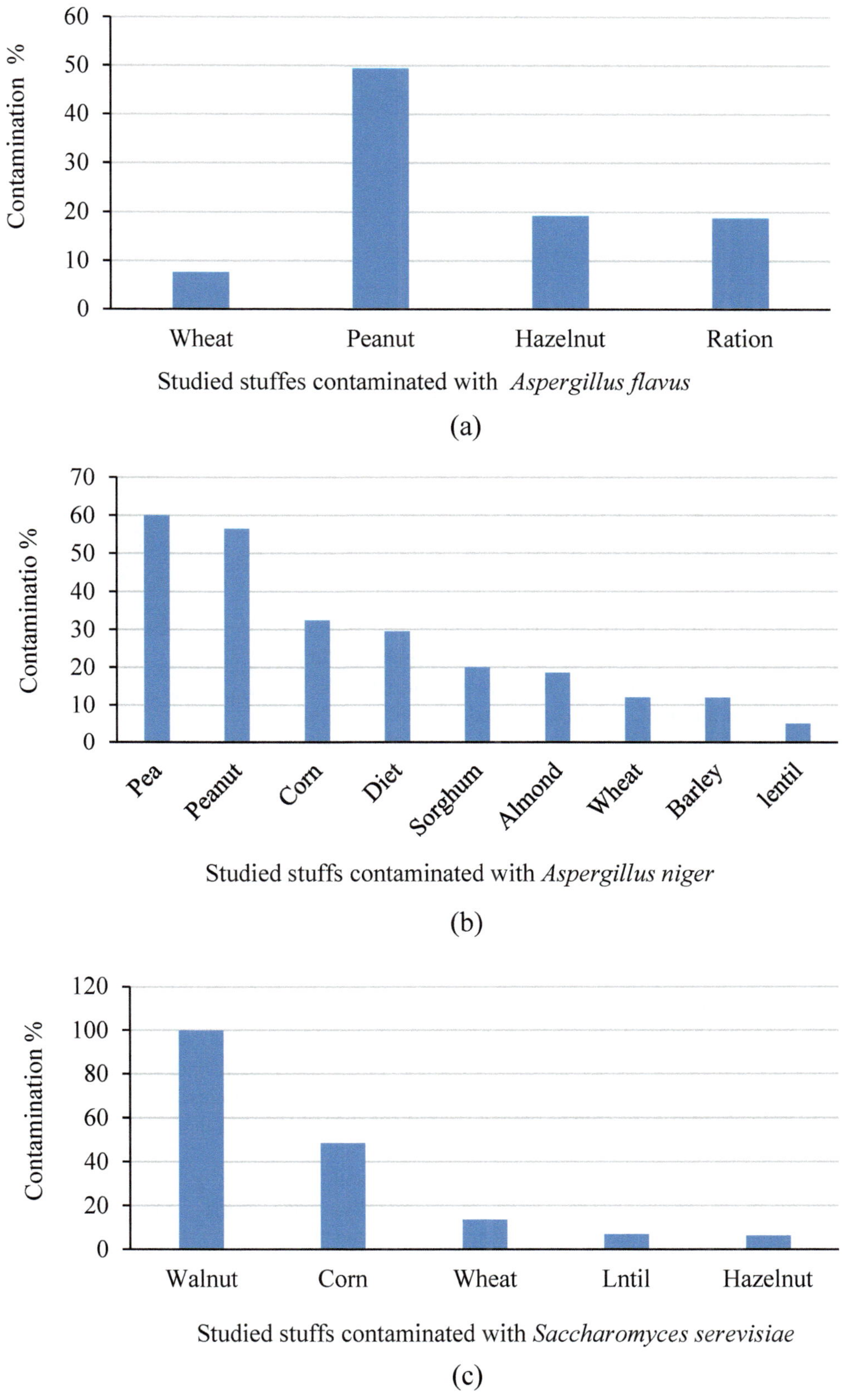

Fig. 7.2 Ranking the studied stuff according to their contamination: with **a** *Aspergillus flavus,* **b** *Aspergillus niger*, and **c** *Saccharomyces cerevisiae* (adapted from: Attitalla et al. 2010a, b)

Table 7.4 Effect of different concentrations of fungicides on the growth of *Fusarium solani* (Abdallah et al. 2005b)

Fungicide	Concentrations ppm					
	25	50	100	200	400	800
Rizolex	9.00	2.50	1.16	0.00	0.00	0.00
Cabtan	9.00	9.00	2.50	1.40	0.00	0.00
Benlate	9.00	1.14	1.00	0.86	0.00	0.00
Roveral,	9.00	3.90	2.6	1.66	1.50	0.00
Roveral,	9.00	7.66	6.75	2.66	0.67	0.00

of incubation at a concentration between 200 and 800 ppm (Abdallah et al. 2005b).

The effect of biological and chemical control on degree of infection and percentage of wilt disease of *Fusarium* wilt (*Fusarium oxysporum*) on two tomato cultivars appeared that the degree of infection in the non-treated plants increased by 72.0 and 73.6%, while in the treated ones it reduced to 24.4 and 2.1% (Buhidma et al. 2007).

The biological control using *Trichoderma viride* and *Aspergillus* sp. was carried out in vitro on three fungi: *Fusarium solani, Alternaria alternate,* and *Stemphylium botryosum*. The results show a decrease in growth length of pathogenic fungi due to the presence of antagonistic fungi. The biological control using the mycorrhizae (A: *Glomus intraradices* and B: *G. etunicatum*) and *Rhizobium* sp. bacteria revealed general reduction in death ratios of seed before and after germination and also reduction in the infection rates. In the green house condition, using the same biological control agents leads to increase in the length, the fresh and dry weight of vegetative and root system, and the number of root nodules (Abdulla 2007).

Trichoderma sp. are widely distributed and nearly present in all soils and other natural habitats, especially those containing well-decomposed organic matter. Soils with very low organic matter are not favorable for the development of *Trichoderma* (Shaban 1986). The investigation on the fungal flora with special attention to *Trichoderma* revealed that *Trichoderma* occurred in moderate frequency in the soil and was isolated from five samples among 23 soil samples collected from Al Jabal Al Akhdar region. The isolated *Trichoderma* counted 0.5–1 $\times$ 10^3 CFU g^{-1} dry soil and identified as *Trichoderma harzianum*. *Aspergillus* and *Penicillium* sp. were the most frequent fungi isolated from the tested soil and were averaged 8.3 and 5.5 CFU mg^{-1} soil, respectively. Several other genera (*Mucor, Fusarium, Alternaria*, and *Curvularia*) were also isolated in moderate, low, or rare frequencies on the studied culture media. *Trichoderma* can be identified by distinctive morphological characters such as rapid growth, bright green, or white conidial pigments, and a repetitively branched (Fig. 7.3). Chlamydospores are usually present and are often abundant, especially in submerged mycelium (Fig. 7.4). Such

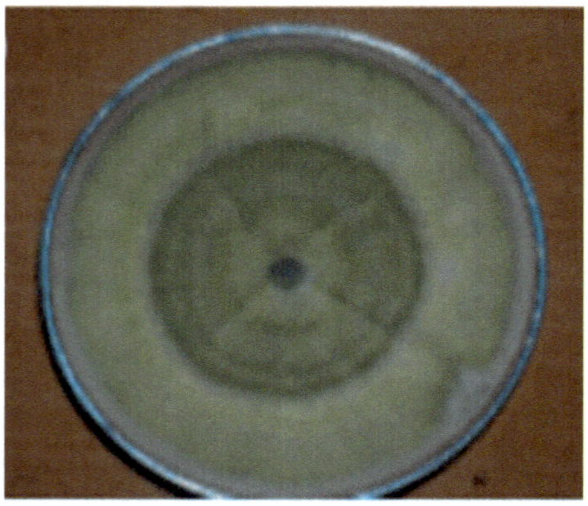

Fig. 7.3 Growth of *Trichoderma harzianum* isolates on PDA plate (Attitalla et al. 2012)

Trichoderma species can be found in different regions in Libya, where the importance of such species as biocontrol agent should be tested for its capacity to control plant diseases (Attitalla et al. 2012).

Several studies and research were carried out to isolate various types of fungi from Libyan soil; for example, *Keratinophilic* fungi are environmental worldwide fungi that degrade keratinous substrates. Assessment of the frequency of *Keratinophilic* fungi, potentially pathogenic fungi from different areas, was the first of its kind in the south of Libya (Altayyar et al. 2016). A total of 136 strains of different *Keratinophilic* fungi were isolated and classified into 10 species belonging to six genera. *Aspergillus* species represent the highest occurrence of isolates.

7.4.1 Mycorrhiza and Truffle

Mycorrhizas in some soils from the coastal area of Libya had been reported, and levels of infection estimated in field samples and the species caused the infections to be characterized as far as possible. Spores of the mycorrhizal endophytes, which were extracted from the soil sampled from plant roots and soil collected in 1973 and 1974 from 14 sites around Tripoli, revealed the presence of mycorrhizas in some soils (El-Giahmi et al. 1976). Thereafter, the studies of mycorrhiza paid more attention to isolating and identifying

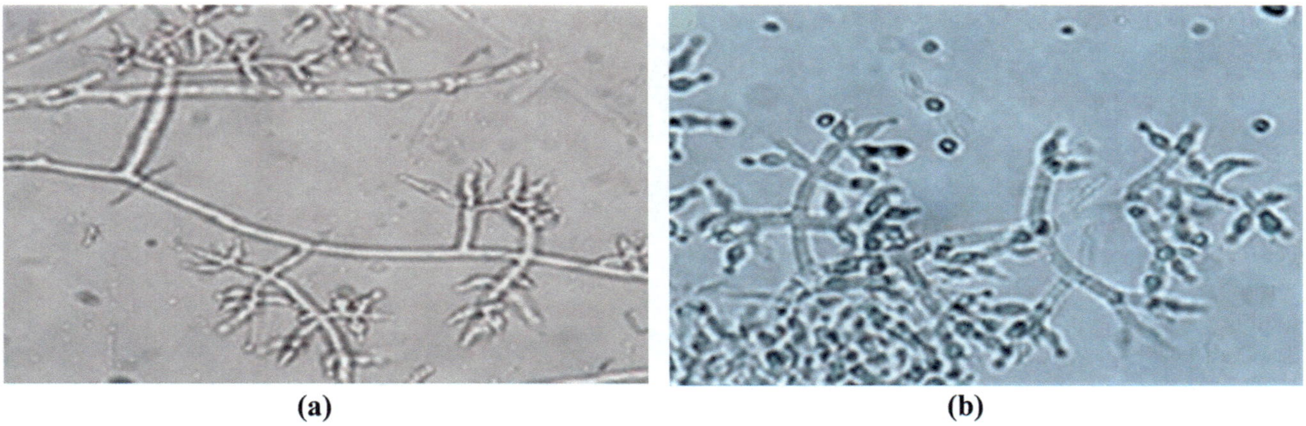

Fig. 7.4 Conidia, conidiophore branching and phialospores of two strains of *Trichoderma harzianum* (**a** and **b**) isolates (Attitalla et al. 2012)

more species. This attention expanded to use some molecular tools to define and classify mycorrhizas. Arbuscular mycorrhizal fungi (AMF) are omnipresent soil inhabitants that are able to form symbiotic associations with root systems of most plant species.

In order to investigate occurrence and distribution of AMF in the Al Jabal Al Akhthar area, east of Libya, roots and rhizosphere soil of 49 cultivated plants belonging to seven families from eight locations were collected. All examined plants were colonized by AMF, and spores abundance ranged from 121 to 992 spores/100 g soil. Spores extracted from soil morphologically identified to genus *Glomus* was the most abundant genus in all studied locations, followed by *Acaulospora, Gigaspora* and *Scutellospora* (AL Ani and Mehjin 2016).

Biodiversity (phenotypic and phylogenetic) of AMF recovered from *Juniperus phoenicea L.* and *Rhamnus lycioides* L. rhizosphere in Al Jabal Al Akhdar was investigated. Spores recovered from soils ranged between 6 and 169 spores g-1soil. The morphology and phylogenetic analysis of 21 sequences of the internal transcribed spacer (ITS) rDNA out of 31 isolated AMF, related them to 11 undescribed species affiliating to eight genera submitted to the GenBank nucleotide sequence under accession numbers from MF599206 to MF599226. A maximum relative abundance (RA) corresponding to 38.7% was recorded for *Claroideoglomus etunicatum*, followed by *Gigaspora* sp. and *Entrophospora infrequens* with RA of 12.9%. *Entrophospora infrequens, Glomus epigaeus*, and *Rhizophagus intraradices* were least encountered in all locations with low RA 3.2%. Phylogenetic analysis results highlighted the dominance of *Claroideoglomus etunicatum* and *Gigaspora* sp. colonizing both *R. lycioides* and *J. phoenicea* in the examined locations (Jadallah et al. 2017).

A study on indigenous mycorrhizal spore was performed in Benghazi region (east of Libya) to determine the presence of AMF spores in a field cultivated with wheat crops (*Triticum aestivum* L.) during the vegetative stage. Three genera of AMF spores were identified (Fig. 7.5). These genera were *Gigaspora* spp., *Acaulospora* spp., and *Glomus* spp. Results

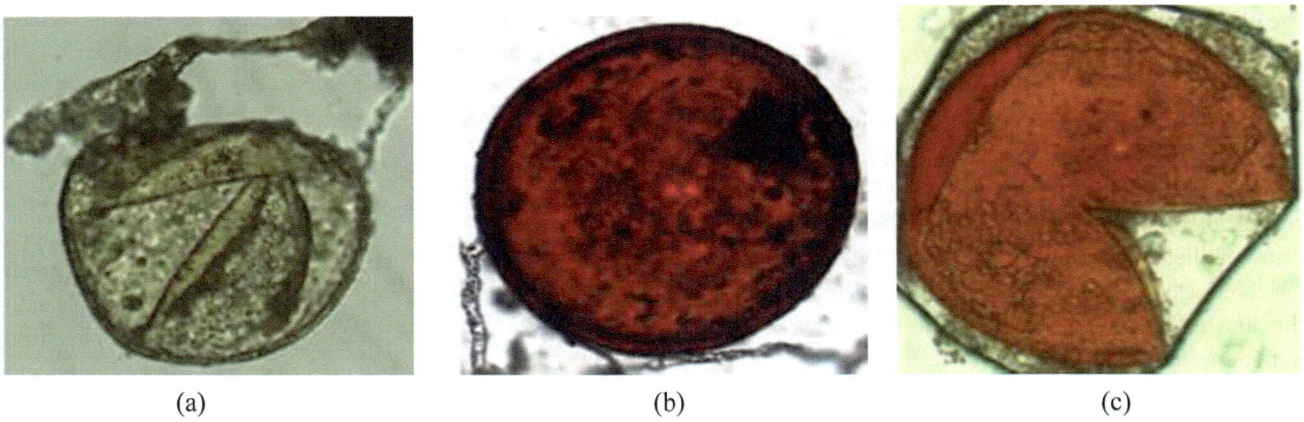

Fig. 7.5 Microscopic pictures of AMF spores collected from cultivated wheat field under light compound microscope (40×). Acaulospora spp. (**a**), Gigaspora spp. (**b**), and Glomus spp. (**c**). Marei (2019)

Fig. 7.6 Partially emerging ascoma causing crack of soil beside the host plant "*Helianthemum* sp." (Moubasher 2010)

showed that *Glomus* spp. was the predominate genus compared to the other two genera (Marei 2019).

Another type of fungus known as desert wild truffles, which are mostly endemic to arid and semi-arid areas, grow naturally in many parts of Libya, especially in Hamada Al Hamra (west region). The development of desert truffle is linked to the amount of rainfall during the winter. The truffle is composed of an ascoma (the edible part) which develops underground and emerges gradually upwards to cause distinct bulging and crack of soil surface (Fig. 7.6). *Helianthemum* sp. are the dominant plant species in the truffle habitat areas, which could have a symbiotic association with the Libyan wild truffles. Locally, truffles are known as Terfas, and in Libya they are very popular and are highly appreciated as traditional food, due to their unique and distinctive taste.

The most known truffle species in Libya are: *Terfezia boudieri* and *Tirmania nivea*. The nutritional values for both genera (*Terfezia* and *Tirmania*) have been proven years ago. Analyses of representative samples of wild truffles—*Terfezia boudieri Chatin* revealed that tubers could be nutritionally considered a good source for protein, iron, potassium, calcium, and sodium. The results indicated that protein was of high quality, since it contained nine essential amino acids, which represents about 6% of the dry weight. Toxic compounds were found to be absent (Ashour et al. 1981).

In a more extensive study of truffles, two kinds of wild desert truffles were collected from Al Hamada Al Hamra. The truffle samples were identified as *Terfezia* (known as red or black truffle) and *Tirmania* (known as white truffle) (Fig. 7.7). Spores collected from some samples are shown in Fig. 7.8. The nutritional values of both (*Tirmania* and *Terfezia*), determined on a dry weight basis, were 16.3 and 18.5% protein, 6.2 and 5.9% lipid, and 67.2 and 65% carbohydrate, respectively. The plants *Helianthemum kahiricum* and *10-Helianthemum lippii* were the dominant plants in the Hamada Al Hamra region found to form a mycorrhiza with desert truffles (Fig. 7.9). The phylogenetic analysis of the genomic rDNA ITS region showed that, out of five collections, three represented *Tirmania pinoyi (Maire) Malencon*, one *Tirmania nivea* (Desf.) Trappe, and one *Terfezia boudieri Chatin*. The sequences were submitted to EMBL nucleotide sequence database with the accession numbers indicated in Table 7.5 (Bouzadi et al. 2017).

7.4.2 Effect of Polluted Soil on Fungi Species and Numbers

It is not only the isolation and identification of fungi in the Libyan soils that has attracted the attention of researchers but also a study of the effect of some soil-contaminated agents on the types and numbers of fungi in areas where pollution is widespread, such as in industrial areas. Physicochemical characteristics of soils in and around Zliten cement factory (Zliten town: 160 km east of Tripoli) have indicated a strong influence by this factory because dusts from the factory have settled on the surrounding soil. Different soil localities support the growth of three fungal genera: *Aspergillus* sp., *Rhizopus* sp., and *Fusarium* sp. isolated. *Aspergillus* sp. has the greatest diversity among fungal species in soil polluted by cement (five species have been isolated). The dominance of fungi of soil was lowest at 100 m from the factory. However, the evenness and diversity increased at this site compared to 300 m and in the control area, which is 100 km away from the factory (Mlitan et al. 2013).

A similar study in an area where an iron and steel factory is located in Misurata (east of Tripoli) proved that all soil

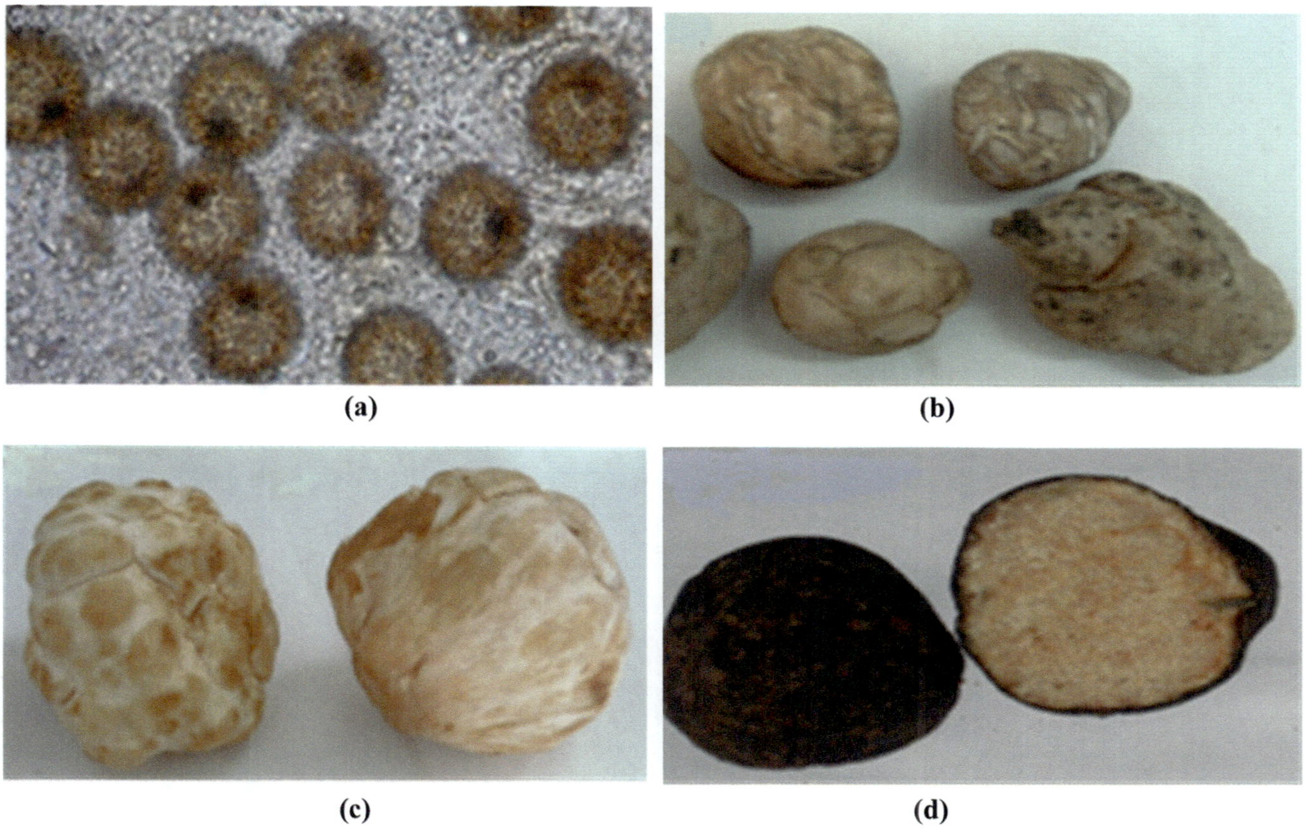

Fig. 7.7 Comparative morphology of collected *Tirmania* and *Terfezia* species: **a** Spores of *Terfezia boudieri*, **b** *Terfezia boudieri*, **c** *Tirmania nivea*, and **d** *Terfezia claveryi* (Bouzadi et al. 2017)

samples exposed to pollution recorded high concentration of heavy metals and were the poorest in fungal population and diversity. Four fungal genera were isolated *Aspergillus*, *Penicillium, Rhizopus*, and *Fusarium*, of which *Fusarium* is the most abundant, and may consider one of the resistance fungi to industrial dusts due to its large spores numbers isolated from polluted area (Mlitan 2013). Another study revealed that the sites near Misurata were affected by industrial waste water pollution. The isolated microbial flora consists of four fungal genera belonging to *Aspergillus*, *Penicillium, Rhizopus*, and *Fusarium*. *Fusarium* and one of the *Aspergillus* species (*Aspergillus sp3*) may influence resistance fungi to industrial waste water (Mlitan et al. 2015).

There are many publications and references issued for the inventory and classification of fungi in Libyan soil. For example: a list of fungi, arranged alphabetically in their classes, followed by a host and substrate index, was published in 1965 under the title "A list of plant pathogenic and other fungi of Cyrenaica (Libya)."

The list also includes notes on the general geography, vegetation, and cultivated crops of the region (Kran and Jürgen 1965). There is also a reference study with a more recent history, which includes a reference list of Libyan fungi titled "Check List of Libyan Fungi." This list was one of the University of Tripoli's publications for the year 1981 (El-Buni and Rattan 1981). Furthermore, a critical checklist of novel fungal taxa having original localities in the arid Middle East region introduced between 1940 and 2000 is presented. The list considers 246 taxonomic units at the species and subspecies levels. The taxa are grouped according to major classes of fungi. For each taxon, the reference, country of origin, and nature of the original substratum are provided (Mouchacca 2005). The checklist and the accompanying bibliography are intended to stimulate the taxonomic research on the mycobiota of this particular arid region.

7.5 Benefits and Limitations of Biological Nitrogen Fixation in Libyan Soil

7.5.1 Pastoral Legumes Scattered in Arid and Semi-arid Regions of Libya

Agricultural and pastoral lands are limited, especially in arid and semi-arid regions that suffer from various desertification factors such as drought, heat, and salinity. Leguminous plants in Libya are estimated at approximately 208 species and are considered to be the third family in terms of the

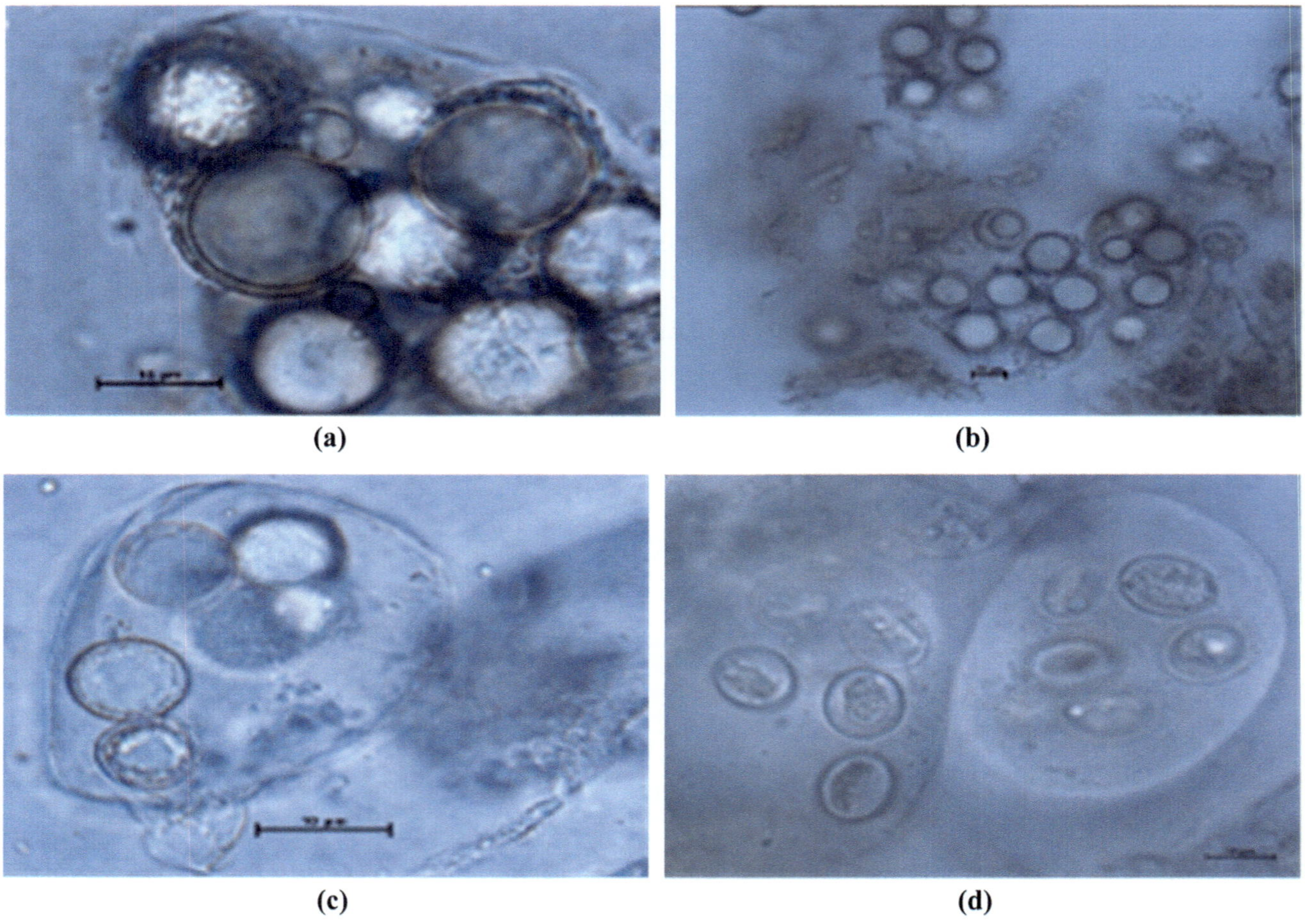

Fig. 7.8 Globose spores from different samples are shown with oil droplets in the center (light microscopy, bar 10 μm), confusedly arranged in ascus. a, b, and c are smooth to nearly reticulate, d is irregularly arranged in ascus. The surface is smooth probably due to unripe stage of spores (Bouzadi et al. 2017)

number of species (Al-Idrissi et al. 1996). These plants are distributed in four main regions: Al Jabal Al Akhdar, the coastal belt, the Fezzan region (especially in some mountain valleys such as Al Hasawneh Al Harouj, Tebasti), and Ghat (see Fig. 2.1, Chap. 2). These leguminous plants do not form natural forests but rather are dispersed in valleys and on the foothills of the mountains in the form of annuals and a few perennials.

Astragalus, Medicago, Trigonella, Lupinus, Retama, and *Acacia* plants are among the most widespread leguminous plants and are distributed in the mentioned areas. These plants, like legumes and other wild plants, play an important role in the Libyan environment, represented by providing other non-leguminous plants with nitrogen, maintaining biological diversity in the Libyan desert, stabilizing sand dunes, protecting soil from erosion and corrosion, and preserving the vegetation of the lands in addition to its use as fodder and pastoral crops for livestock by herders (Atallah et al. 2008).

Acacia species, which are members of the Leguminosae family, can grow on barren soils and dry sites that are not suitable for many plant crops. *Acacia* are widespread and cultivated throughout the country, mainly for stabilizing soils and building wind-breaks. *Acaci tortilis* (*subsp. raddiana*) grows naturally in arid areas at south of Libya (Fezzan, Wadi-Eshati and Soukna). It is used as fodder for goats and camels in areas where natural grazing vegetation is rare. *A. cyclops* is restricted to the Mediterranean Sea coast (north of Libya), growing in the areas subjected to sea winds. Meanwhile, *Acacia karroo* is cultivated around the farms to form an impenetrable barrier to livestock. *Acacia cyclops* are also common in Libya, and other *Acacia* spp. such as *Acacia nilotica, Faidherbia albida*, and *Acacia farensiana* can occur, but in limited numbers (Jafri 1978). Another advantage of the *acacia* that encourages its cultivation is that it improves the properties of the soil and raises its fertility due to its ability to enter into symbiosis with rhizobia and vesicular arbuscular mycorrhizal (VAM) fungi.

(a)

(b)

Fig. 7.9 **a** Al Hamada Al Hamra habitat areas of the collected *Tirmania* and Terfezia *species* and the potential host plant species. **b** *Terfezia boudieri* under their hot plant *H. sessiliflorum* (*H. lippii var. sessiliflorum*) (Bouzadi et al. 2017)

Fezzan is a part of the region that occupies a large area of southern Libya where wild pastures are spread, such as wild grass and pastoral legumes (Fig. 7.10). In order to visualize the natural image of Fezzan, the region must be described from the climatic and biological aspects to set expectations for the type of life in this region. Fezzan is a very large area of sand interspersed with some dry mountains and valleys, with scattered oases in which Libyans are living and which depend upon ground water. The desert climate is prevalent, and there is a significant difference between the temperatures. In the summer, sometimes it can reach 46 °C, while in winter, it drops to less than zero. Rainfall is scarce and does

Table 7.5 List of analyzed samples with basic information, herbarium voucher numbers, identification, and GenBank accession number. *Tirmania* 1, 2, 3, and 4 collected from lower region of Al Hamada Al Hamra; meanwhile *Terfezia* from upper region of Al Hamada Al Hamra (north-west of Libya) (Bouzadi et al. 2017)

Herbarium voucher at Herbarium Juva Truffle Centre[a]	DNA voucher deposited at the Slovenian Forestry Institute[a]	EMBL nucleotide sequence database accession numbers
Tirmania 1-2008/HHU	E09_4_13	FN395012
Tirmania 2-2008/HHU	E09_4_14	FN395013
Tirmania 3-2008/HHU	E09_4_15	FN395014
Tirmania 4-2008/HHU	E09_4_23	FN395015
Terfezia 5-2008/HHL	E09_4_22	FN395016

[a]The samples (ascocarps) were deposited in Herbarium of Juva Truffle Centre, Finland and parallelly in the Herbarium and Mycotheca at the Slovenian Forestry Institute (ascocarps and extracted DNA)

Fig. 7.10 Wild pastures of wild grass and pastoral legumes spread in large area in the south of Libya (Khalifa 2013)

not exceed the annual average of 20 mm. Groundwater is the main source for both drinking and irrigation systems. Environmental stresses like high summer temperature, salinity, drought, low soil fertility, and wind are the major challenges that Libyan farmers face (Zurqani et al. 2019). Nevertheless, this large area of land was placed into cultivation, and different legumes and cereal crops are cultivated in the oases. Cowpea is one of the earliest legume crops known to the inhabitants of Fezzan. It has gained a wide popularity between farmers in many oases because it is used mainly as a source of food. Although Libyan flora is rich in leguminous plants and occupies the third position among other families, studies related to the description and characterization of symbiotic rhizobia with these plants are still limited and have been limited to some cultivated crops and woody legumes (Mohamed et al. 2000; El Sadi 2004), as well as some types of wild clover (*Medicago* spp.) (Mohamed 2018).

Many Libyan wild plants are threatened with extinction due to drought, salinity, and overgrazing (Zurqani et al. 2018). That is why the interest in these plants in terms of preserving their seeds and their propagation was urgent and received some attention from those interested in agriculture and the environment. However, research on endemic rhizobia that is compatible with these plants has not been provided with adequate attention, and information about their physiological properties and their symbiotic features is still limited. The most important goals that researchers sought in their studies and research include: searching for symbiotic rhizobia with some wild leguminous plants that are abundant in arid and semi-arid lands from Libya; testing the ability of isolates to symbiosis and nitrogen fixation with cultivated leguminous plants; testing the ability of these isolates to withstand environmental and chemical stress; studying phenotypic variation; and comparing it with isolated reference strains from different geographical regions. These goals were chosen in order to obtain strains resistant to various environmental factors that hinder the symbiotic process and the biological fixation of nitrogen.

7.5.2 Challenges of the Symbiotic Nitrogen Fixation

The tendency to use biofertilizer by root nodules bacteria instead of chemical fertilization has had a positive impact on the environment, especially in arid and semi-arid areas. Inoculation of *Acacia saligna* by four different strains of *Rhizobium* sp. under different levels of salt stress revealed that the inoculation of plants significantly reduced the harmful effects of saline and increased the ability of inoculated plants in their tolerant different concentrations of salinity (Zurqani et al. 2018). Furthermore, number of root nodules formed for each plant, the dry weight of root-nodule, and the total concentration of nitrogen in the soil significantly increased in inoculated plants compared with the control (Table 7.6) (Mohamed 2018).

Despite the importance of these *Acacia* species in Libya, a little attention has been given to their root-nodule symbionts. Libya has long hot summers and short rainfall periods, and its soils are affected by salinization and drought. For this there was an intent study to investigate the symbiotic and phenotypic characteristics of *Acacia* spp. root-nodulating bacteria in order to select effective and competitive strains tolerant to different climatic conditions. Three different regions of Libya were selected for the study by Mohamed et al. (2000): a Saharan area (Marzuq) where all *Acacia* spp. are cultivated except *A. tortilis* (subsp. *raddiana*) and two semi-arid areas along the Mediterranean, the east (Benghazi) and the west (Tripoli). A total of 30 root-nodulating bacteria were isolated from five different *Acacia* species. No nodules were found on *A. cyanophylla* and *A. tortilis* in the Saharan region (Marzuq). However, nodules were obtained from their rhizosphere soil in the laboratory. A numerical analysis of 104 characteristics formed five distinct clusters. Most clusters grew at temperatures 35 and 37 °C; some grew at temperatures above 40 °C. Regarding tolerance to NaCl on agar medium, the majority of isolates were unable to grow at a concentration of 2% NaCl. Some, though, were highly resistant, and there was one isolate that even grew at 8% NaCl.

Rhizobial isolates were recovered from root nodules of two cowpea cultivars that grow in four different sites in Fezzan (Abdelnaby et al., 2015). The study of symbiotic and phenotypic characteristics of rhizobial isolates revealed that all isolates formed symbiosis with the tested plants, but they were different in their nitrogen-fixation efficiency. Numerical analysis of phenotypic characteristics of the isolates formed four distinct groups (Fig. 7.11). Most isolates exhibited wide tolerance to acidity, alkalinity, and extreme temperatures. Isolates displayed different responses to salinity ranging from sensitive, which are unable to grow in 1% NaCl, to resistant, which grow at 2% NaCl or above. The four distinct phenotypic groups formed could indicate the genetic diversity of the isolates and different species may be present in the soils of Fezzan (Abdelnaby et al. 2015).

Phenotypic studies are useful for the selection of superior rhizobial strains adapted to environmental stresses such as salinity, acidity, and elevated temperatures. Such studies could increase the knowledge about the symbiosis of rhizobia and help us select superior isolates for use as inoculum. On the other hand, another study in different regions of Libya (north-east and north-west) intended to select native isolates that may be able to withstand the local environmental conditions of Libya and enhance nitrogen fixation with cultivated forage legumes such as *Medicago sativa*. Indigenous rhizobial isolates of root nodules of wild legumes (*Trigonella stellate* L. and *Medicago polymorpha* L.) were effective in their symbiosis with their cross-inoculation *Medicago sativa* L. The numerical taxonomic analysis showed a similarity level of 87% and formed four distinct groups. The physiological characteristics of the isolates demonstrated a wide diversity in their tolerance to salinity, from sensitive, which were unable to grow at 1%, to

Table 7.6 The effect of inoculation with different strains of rhizobia under different levels of salinity on nitrogen concentration in the soil (mg/kg). NA0 is control (uninoculated). NA1, NA2, NA3, and NA4 are different strains of Rhizobia. NA4 was the superior strain (Mohamed 2018)

Strains of bacteria	Salt level ds m^{-1}			
	4	5	8	10
NA0	12.5	10.11	6.25	1.52
NA1	15.42	13.52	7.4	3.22
NA2	18.22	14.65	10.22	5.21
NA3	22.10	17.32	13.33	6.42
NA4	25.33	20.11	15.1	9.08

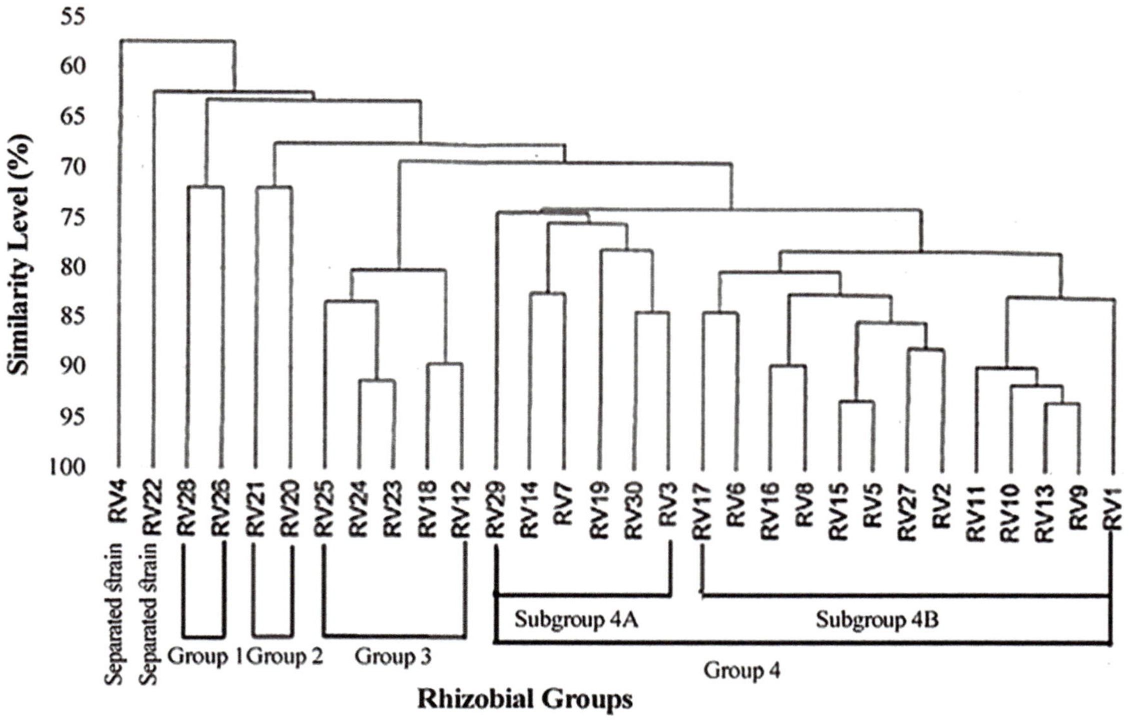

Fig. 7.11 Dendrogram showing the phenotypic relationship between the tested isolates. Four groups were formed from the numerical analysis with 70% similarity (Abdelnaby et al. 2015). Group 1: Seemed effectively nodulated their hosts and *Arachis* plant, but ineffective with *F. albida.* Group 2: Formed ineffective symbiosis with all test plants including their hosts. Group 3: Included five isolates which formed indigenous cowpea, appeared ineffective in nodules formation on the test plants (*A. hypogaea* and *F. albida*). Group 4: All of them formed ineffective nodules on *F. albida*, but different in their symbiosis with their hosts and *A. hypogaea*

resistant, to high salinity, which are able to grow at 4% of NaCl. Additionally, most of isolates from both plants were highly tolerant to high temperature and grew at 44 °C; one strain showed the greatest tolerance to high temperature at 48 °C (Mohamed et al. 2018a).

The symbiosis between rhizobia and leguminous plants is a sensitive process for environmental factors, especially saline stress, as it prevents the first step of the symbiosis process (Zahran 1991). Rhizobia can tolerate salinity more than the host plant can, but their degree of tolerance varies (Zahran 1999). Inoculation of leguminous plants with saline resistant strains of rhizobia may help the host plant to resist salinity. From this standpoint, it became necessary to search for saline-tolerant rhizobial isolates, grow legumes in arid and semi-arid regions, and select the isolates that have the potential to improve symbiotic process and nitrogen fixation in saline soils.

In order to determine the level of sodium chloride salt that allows the formation of symbiosis and nitrogen fixation, a survey of 84 isolates were isolated from different legumes, planted or wild, in different locations of the dry zone (Fezzan). Three groups were formed when the results of the salinity test were subjected to numerical analysis. Most of the isolates in the first group were resistant to all tested salts. These isolates were from non-saline soils and grew at a salinity level of 8.5–9.5% on agar plates. The isolates of the second group, which was isolated from saline soil, were sensitive to sodium and magnesium chloride salts. The third group of isolates varied in their response to salts tested. Meanwhile, the effect of NaCl salt on the symbiosis between alfalfa plant and some isolates from the fenugreek showed that the maximum concentration allowed for the formation of symbiosis was between 0.8 and 1% of NaCl. 1.5% level was a disincentive to the symbiosis process (Mohamed et al. 2018b).

More exclusive investigation of the symbiotic rhizobia was studied with some wild plants in different regions of Libya, including arid and semi-arid regions, specifically in the north-west (Mazda, Gharyan, Yafran, Nalut, Darj, and Ghadames), in the north-east (Ras Al Hilal and Al Bayda), and in the south (Ghat, Wadi Al Shati, and Jabal Al Hasawnah) (see Fig. 1.1, Chap. 1). Phenotypic variety was studied for 51 rhizobial isolates isolated from the root nodules of five species of wild legumes: *Trigonella stellata*, *Medicago polymorpha* L., *Astragalus* spp. Bioss, and *Lupinus varius* L. (Khalifa 2013). The symbiotic features of the rhizobial isolates showed that the isolates from the *Astragalus* spp. and *Lupinus varius* (Fig. 7.12) were specialized and were symbiotic only with their hosts (Fig. 7.13). Meanwhile, the isolates of *M. polymorpha* and the wild

(a)

(b)

Fig. 7.12 *Astragalus fructicus* Bioss (**a**) and *Lupinus varius* L. (**b**), grown in sterile soil in plastic pots, seed sterilization method, lighting conditions, and temperature were similar to those used in seedling germination in test tubes (testing of the isolated strains of rhizobia from the host plant) (Khalifa 2013)

fenugreek *T. stellata* formed an effective symbiosis with the cultivated *M. sativa* (Fig. 7.14). Nodules of *Lupinus varius* are shown in Fig. 7.15. Twelve groups, regardless of their host plant or place of isolation, were formed from the numerical analysis. Two groups of these isolates can be selected for use as an inoculum for (*Lupinus* spp.) in land affected by salinity, while isolates of some other groups, such as alfalfa, can be used for forage crops in lands less affected by salinity. The results of the physiological characteristics of isolates showed that they were distinguished in their tolerance to the environmental conditions from which they were isolated. Most of them grew at a temperature above 40 °C, and most isolates of *Lupine varius* were extreme in their tolerance, growing even at temperatures above 47 °C. The tolerance of isolates to salinity differed between sensitive, which does not grow in a concentration of 1%, to an extreme in its resistance, which grows up to 9% (Fig. 7.16).

The *Melilotus officinalis* plant is one of the pastoral leguminous plants (Fig. 7.17) and is similar in appearance to the *Medicago sativa* (Fig. 7.18). It is endemic and has spread throughout the country. The plant is considered drought-tolerant but is very sensitive to salinity. *Melilotus* sp. plants have a symbiotic relationship with rhizobia, but there is no sufficient information about the biochemical and physiological properties and their symbiotic characteristics. For this reason, an investigation was adopted to study the physiological properties of pure isolates of bacteria that coexist symbiotically with the *M. officinalis* plant grown in Fezzan. Two groups were obtained from the numerical taxonomy of isolates at a similarity level of 81%. The results of the physiological characteristics showed differences between the groups of isolates. With the exception of sensitivity to most of the tested heavy metal toxins, most isolates adapted to the tolerance to NaCl, from sensitive (unable to grow at 1% NaCl) to extremely tolerant (growing up to 5%.) Most of the

Fig. 7.13 Bacterial nodules forming on the roots of its host plant *Astragalus fructicus* Bioss (Khalifa 2013)

Fig. 7.14 The test of *M. polymorpha* strain inoculum on *M. sativa* compared with non-inoculated plant (control). Small white nodules formed on the roots of *M. sativa* but did not form on the non-inoculate plant. In addition, the plant yellowed and wilted due to nitrogen deficiency in the non-inoculated plant (Khalifa 2013)

isolates grew at temperatures higher than 44 °C and tolerated high degrees of alkalinity (exceeding pH 9). The isolates distinguished in their tolerance to various stress factors could be nominated for use as inoculum for forage crops such as alfalfa (*M. sativa*), and cultivated legumes such as fenugreek (*T. foenumgraecum*), due to its high competitiveness (Ibrahim et al. 2019).

Sea water is a major source of salts that reaches coastal agricultural lands through the drizzle (Aerosol) or through interference with groundwater that leads to what is known as secondary salting as a result of the accumulation of salts. Determination of the tolerance of *Rhizobium leguminosarum* isolates to different concentrations of salinity stress, which were isolated from faba bean grown on different soil types of Al Jabal Al Akhdar region, revealed that some variations were observed in bacterial response to different salt stress. The differences in response to salinity reflect the adaptation of the strains to tolerate increased salinity stress and the direct and indirect effects of salts. At high salinity levels, detrimental effects were also observed. The difference in response between the strains may also be attributable to the differences in the chemical and physical properties of the soils from where the strains were isolated. This soil is not saline, but the strains would have been subject to severe salt conditions during droughts periods that characterize the Mediterranean regions, which are located within the semi-arid zone (Shoaib et al. 2009).

Studying the effects of Na^+, Ca^{2+}, Mg^{2+}, and Boron (B) concentrations and sodicity on the growth of *Rhizobium leguminosarum* isolated from a salt-affected soil led to the conclusion that Na^+ concentrations of more than 1.1 g L^{-1} impeded growth, especially at high sodium adsorption ratio (SAR) values. Mg^{2+}, added together with Na^+ or with (Ca^{2+} + Na^+), affected growth more negatively than (Ca^{2+} + Na^+) alone. It was concluded that EC of more than 4 dS m^{-1} retarded the growth of *Rhizobium*, but only at high sodicity levels. When SAR increased from 10 to 30, Na^+ had no clear effect on growth, irrespective of the accompanied cations. Growth was reduced by B concentrations as low as 0.5 mg L^{-1}, and the B effect was enhanced by increased salinity (Faituri et al. 2001).

Fiatori et al. (1996) concluded in a previous study that salinity and soda had a clear negative effect on the symbiotic relationship between bacteria and bean plant and that high soda levels led to a significant increase in the number of root nodules. However, this increase was not accompanied by any increase in the amount of total nitrogen, which may indicate an inefficiency of the root nodules. Therefore, when investing lands affected by salts or when irrigating with salty water, it is necessary to choose the most tolerable varieties of legumes that are salt-tolerant with the use for inoculum of *Rhizobium* bacteria isolated from the same soil for cultivation.

In order to solve the problem of salinity, consideration must be given to the facts of research findings and related recommendations. These studies may explain the presence of rhizobia isolates that are suitable for different soils and for different environmental conditions prevailing in the region.

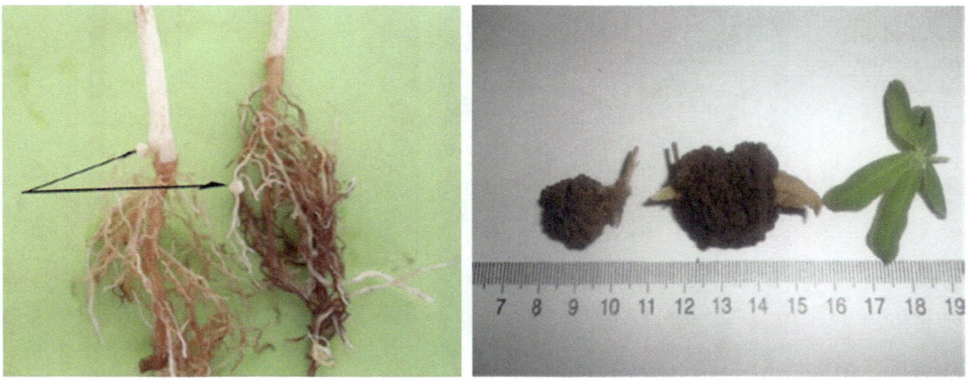

(a) (b)

Fig. 7.15 The root nodules formed on the roots of *Lupinus varius* L. plant that were planted in the laboratory (**a**). The root nodules on the roots of the *Lupinus varius* L. plant that were planted in the field (**b**). It was large (2–3 cm in size) and clustered around the original root (Khalifa 2013)

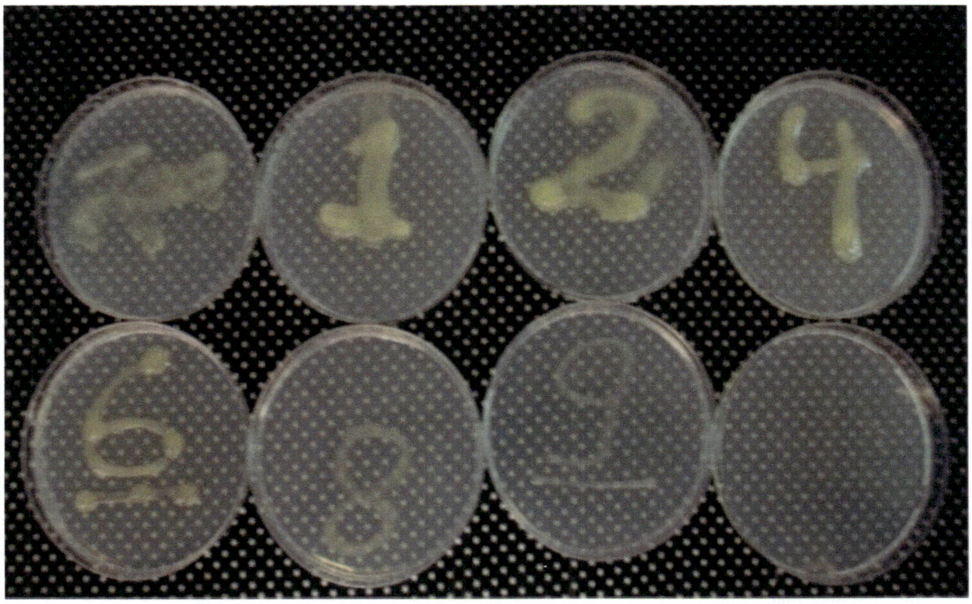

Fig. 7.16 The growth of a *Rhizobium* strain isolated from a *Lupinus varius* L. plant in different concentrations of NaCl. (Khalifa 2013)

(a) (b) (c)

Fig. 7.17 The *Melilotus officinalis* plant appears in the field with its yellow flowers (**a** and **c**), as do bacterial nodules forming on its roots (**b**) (Ibrahim et al. 2019)

Fig. 7.18 Alfalfa plant *Medicago sativa* in the field and its roots after removing it from the soil. (Khalifa et al. 2018)

The isolates of rhizobia from saline regions were not more resistant than isolates from non-saline regions; the resistivity of salinity is not related to plant or isolation site but is considered characteristic associated to rhizobia. Tolerance to salinity can give rhizobial isolates the advantage of endemism and reproduction in saline soils, but the formation of nodules and nitrogen fixation depends on the type of legume plant. The most salt-tolerant plants can have a symbiotic relationship with rhizobia and with nitrogen fixation. Some strains may be candidates for selection to inoculate legumes in saline soils if their ability to fix nitrogen under saline stress and develop salinity tolerant plants is proven. It is noted from the results that the tolerance of isolates to salinity is higher than that which allows the formation of symbiosis, and this is due to the sensitivity of the symbiosis process to saline stress.

Some strains of rhizobia not only have a symbiotic relationship with leguminous plants but also have an ability to produce enzymes. Like any other living organisms, rhizobia must be able to perform a set of chemical reactions to continue to survive, grow, and multiply. These interactions are carried out by enzymes in sequential reactions known as nutritional metabolism.

To find out the ability of some rhizobial isolates isolated from the *Medicago sativa* (Fig. 7.18) to produce some enzymes, a reference strain and different rhizobial isolates were examined to determine their ability to produce a group of enzymes. Isolates varied in their production of enzymes, but all, including the reference strain, were positive to catalase, oxidase, and urease enzymes, but different in the production of other enzymes (Fig. 7.19). Some of the produced enzymes were considered very important to rhizobia, as illustrated in Fig. 7.20. These strains could be candidates for use as an inoculum for leguminous plants in environments that are exposed to different stress factors, especially in arid and semi-arid regions with poor organic matter content (Khalifa et al. 2018).

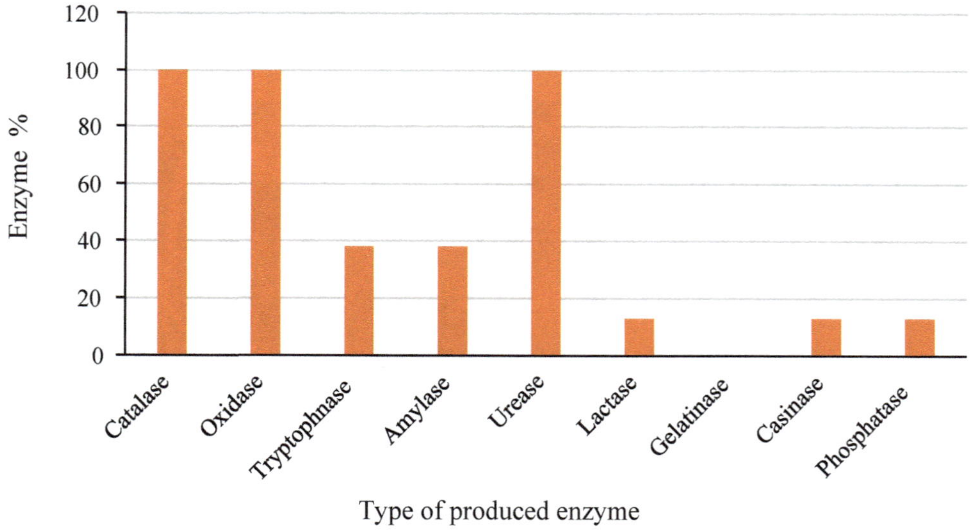

Fig. 7.19 The percentage of different enzymes that were produced by the *Rhizobium* spp. isolates isolated from *Medicago sativa* (modified from Khalifa et al. 2018)

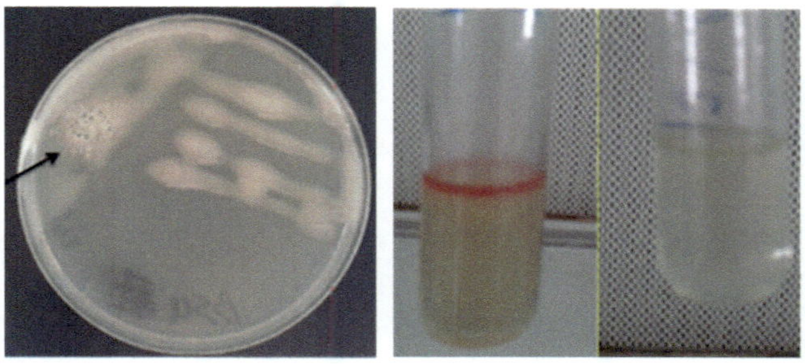

(a) (b)

Fig. 7.20 Reaction of the catalase and tryptophanase enzymes: **a** The emergence of bubbles as an indicator of the production of catalase enzymes after the addition of H_2O_2. Catalase is one of the respiratory enzymes that aerobic bacteria use to get rid of the compound H_2O_2, and **b** The appearance of the red ring is an evidence of the production of tryptophanase enzyme (positive). The enzyme tryptophanase is used by rhizobia to break down the amino acid (tryptophan) into indole, pyruvic acid that is used by bacteria to produce energy, and ammonia exploited by bacteria as a source of nitrogen in the absence of leguminous plant (Khalifa et al. 2018)

7.5.3 Benefits of Non-symbiotic Nitrogen Fixation

The density of *Azotobacter* in the rhizosphere and in root-free soils in different Libyan soils, including some oases and some fields of vegetable crops, were determined. *Azotobacter* was detected in adequate numbers in soils from Tripoli, El-Marj, and Agdabia, and the rhizospheres of cultivated soils contained higher counts of *Azotobacter* than root-free soils. Meanwhile, the oases soils contained none or very few numbers. *Beijerinckia* was found only in the soils of El-Marj, El-Kuffra, and Hoon. The rhizosphere soil of alfalfa had higher counts of *Azotobacter*, followed by broad-bean, wheat, and barley. In the case of soils under vegetable plants, tomato soils had the highest counts of *Azotobacter* compared with carrot soil (Makawi 1973). When the potato cultivar *Arran Banner* inoculated with *Azotobacter chroococcum*, the plant responded significantly, and its growth was stimulated. The yield of marketable tubers increased by 8.5–42.6% (Imam and Badawy 1978).

Azotobacter inoculum can increase the rate of nitrogen fixation with non-leguminous crops, subsequently improving soil fertility. Using *Azotobacter* species isolated from Wadi Elrabie soil (Tajura) as inoculum to inoculate Barely Hordeum had a positive role in increasing the dry weight of the shoot and root of the plant and also led to an increase in the density of *Azotobacter* in the rhizosphere (Ben Mahmoud and Frejani 2016). In another study, inoculation with nitrogen-fixating bacteria, *Burkholderia* ssp. isolated from Libyan soil and different rates of nitrogen fertilizer (ammonium phosphate) increased the productivity of the tomato. The highest percentage of productivity was achieved when

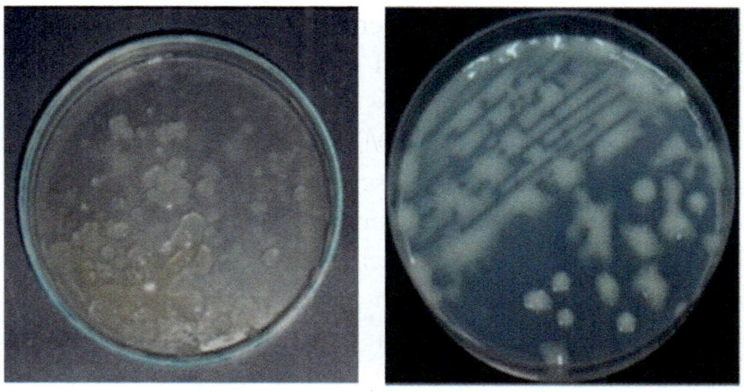

(a) (b)

Fig. 7.21 Colonies of isolated *Azotobacter* sp. (**a** and **b**) on LG medium Incubated at 30 °C for 3–7 days. Microscopical and biochemical tests classified the isolated bacteria as *Azotobacter* sp. (Ben Mahmoud and Ferjani 2018)

Burkholderia sp. inoculum was used with 180 kg N/ha of fertilizer, where the yield was 17.50 kg/plant (280.83 ton/ha) (Ben Mahmoud 2016).

The influence of soil pH on *Azotobacter* population using microbiological characteristics as bio-measurement in arable lands was detected in 15 samples of soil collected from the rhizosphere soil from different regions in Tripoli. The microscopical and biochemical tests classified the isolated bacteria as *Azotobacter* sp. (Fig. 7.21). The results showed that all soil samples contained *Azotobacter*, and the high population of *Azotobacter* was observed in soil samples with the range of pH (7–7.5). *Azotobacter* tended to grow better in neutral soils than in slightly alkaline soils. Increasing the nitrogen percentage in soil corresponds to increasing the *Azotobacter* population (Ben Mahmoud and Ferjani, 2018) (Fig. 7.22).

Inoculation of grass-cycling compost with N_2-fixing bacteria may improve its quality by increasing total nitrogen and available phosphorus. Inoculation of compost with *Azotobacter* sp., *Burkholderia* sp. and *Azospirillum* sp., each alone and all three together, increased the total N by 5–15% (Fig. 7.23), and the available P by 20–30% (Fig. 7.24) in comparison with the uninoculated. Increasing the N content and P availability of compost increases its value, and there may be additional benefits from providing N_2-fixing bacteria (Ben Mahmoud et al. 2018). Some studies dealt with biofertilizer, but they did not reach the point of commercial use and were limited only to laboratories and simple trading.

7.5.4 The Effect of Pesticides on Bacteria Related to Soil Nitrogen

Some pesticides remain in the soil for a long time and resist degradation, causing negative effects on beneficial organisms such as nitrogen-fixing bacteria. Soil microbes are

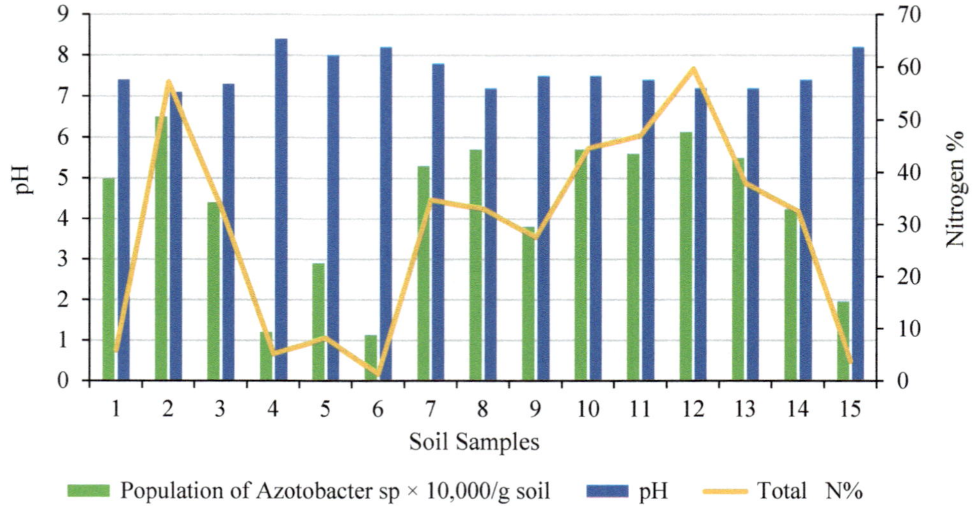

Fig. 7.22 The relationship between pH, *Azotobacter* population 10^4 g^{-1} soil and total nitrogen content (N %) in soil samples. (modified from Ben Mahmoud and Ferjani 2018)

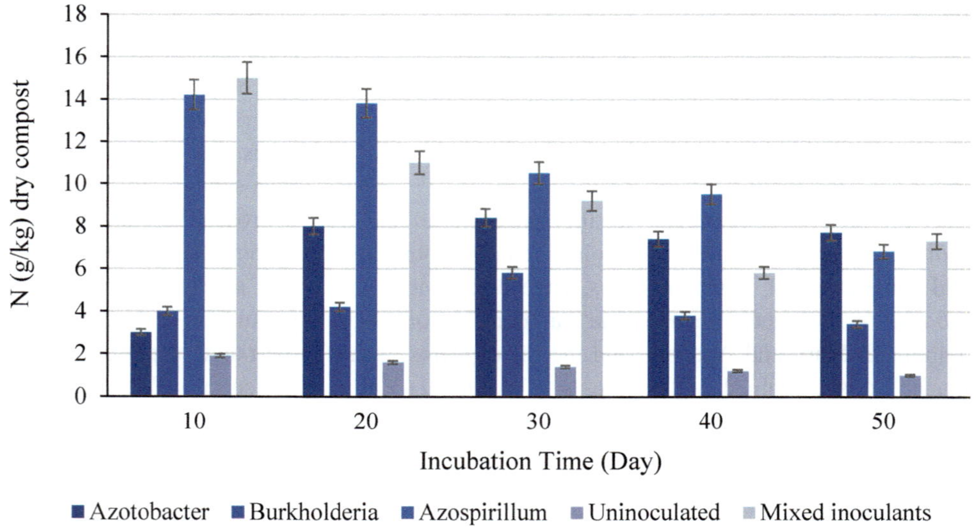

Fig. 7.23 Overall gain in N_2 content of compost from N_2 fixation process. (modified from Ben Mahmoud et al. 2018)

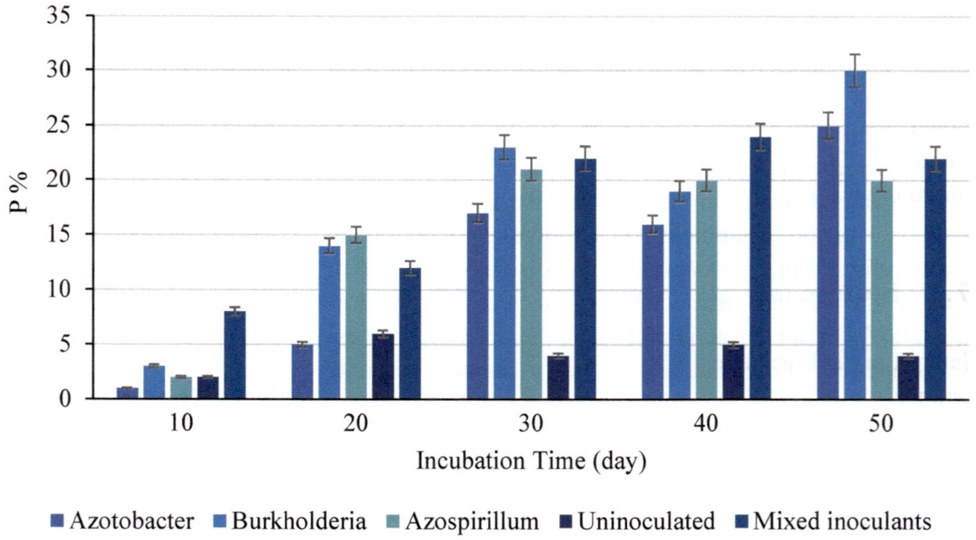

Fig. 7.24 P (%) content of compost with time following inoculation of 40 days old compost with N-fixing bacteria singly and mixed inoculants (modified from Ben Mahmoud and Ferjani 2018)

affected in different degrees by pesticides. The most affected biological process of using pesticides is the symbiotic nitrogen fixation process. Quantitative effects of Temik (insecticide and nematocide), Orthocide (fungicide), or Treflan (herbicide) were tested on soil microorganisms. The effect of any pesticide on counts of *Azotobacter*, N-fixing *Clostridia*, and ammonifiers was subsequently depressive. However, the autotrophic nitrifying bacteria, especially ammonium oxidizers, seemed to be more sensitive, and their counts in treated soil were sharply depressed (Makawi et al., 1979). Zaid et al. (2014) found that *Terbutryn*, *Propyzamide*, and *Propyzamide* had no significant effect on nitrogen-fixing bacteria, except in the case of carbetamide, which increased the total number of nodules per plant. Actinomycetes were reduced by all herbicides treatments, but the total number of soil fungi increased by using *Propyzamide* and *Metribuzin*.

In other cases, Elssaidi and Mohamed (2015) found that *Rhizobium* bacteria that isolated from fenugreek, *R. meliloti*, was the most affected and sensitive species for pesticides. Additionally, not all the rates of pesticides application were effective, and, in the short term, no harmful effects were observed. However, bioaccumulation, in the long term, may lead to chronic toxicity. Another study revealed that the use of pesticides led to a decrease in numbers of *Nitrosomonas*, *Nitrobacter,* and cellulose-decomposing bacteria at the recommended concentrations and dose. Also, there was a highly significant interaction between the type of insecticide and its concentration (Al Hammadi 2015). Using fungicide (Benomyl 50%) and insecticide (Abamectin) had a different effect on the growth of nitrogen fixing bacteria *Azotobacter* and *Burkholderia*, depending on the concentrations and incubation periods. *Azotobacter* and *Burkholderia,* with the dilution of 10^{-4}, were more tolerant of pesticides as compared with other dilutions of 10^{-5} and 10^{-6} (Ben Mahmoud and Ferjani 2017).

The bottom line of these studies is that pesticides can have an inhibitory or stimulating effect on microorganisms in soils, and the period of retardation or stimulation is different according to the chemical composition, the concentration of the pesticides, and the physiological properties of micro-organisms.

7.6 Bioremediation of Crude Oil

Interest in bioremediation field has increased in recent years, and many studies have shown that microorganisms have the potential to be used for the clean-up of sites contaminated by crude oil. Hesnawi and Mogadami (2013) revealed that mesophilic bacteria was more beneficial for biodegradation than thermophilic. The observed total petroleum hydrocarbons (TPHs) removal was 79.8 and 62.4% at mesophilic and thermophilic temperatures, respectively.

A scale study by Shaieb et al. (2015) indicated that four bacterial isolates from the Libyan desert were identified as *Cellulosimicrobium cellulans*, *Brevibacterium liquefaciens*, *Enterococcus saccharolyticus*, and *Brevibacterium mcbrellneri.* The concentration of the TPHs in the cultures of the selected isolates was decreased to 18.86, 26.16 and 39.04% for the first three genera, respectively. The bacteria isolates have the potential to be used in the bioremediation of crude oil. El Mahdi et al. (2015) revealed that 20 crude oil-degrading bacterial strains were isolated from oil-contaminated sites at Al Hariga Oil Terminal and Nafoora Oil Field. Two isolates were identified by 16S rDNA gene sequence analysis to be *Pseudomonas aeruginosa* NAF1 and *Kocuria pastrius* SAR3. *P. aeruginosa* NAF1 exhibited 70 and 76% crude oil degradation. Likewise, *K. pastrius* SAR3 expressed 68 and 70% crude oil degradation in batches supplemented with 0.2% corn steep liquor (CSL) and solid waste date (SWD) within 28 days. The degradation efficiency of *P. aeruginosa* NAF1 also increased to reach 91 and 97% using 0.5% (w/v) (CSL) and (SWD), respectively.

The oil industry generates massive amounts of crude oil tank bottom sludge (COTBS), a complex mixture of hydrocarbons. COTBS contains a significant amount of recoverable oil (42.08%). *Hydrocarbonclastic* bacteria were isolated from COTBS and characterized in terms of their *hydrocarbonoclastic* potential. Three bacterial isolates (*Pseudomonas* sp., *Pseudomonas xanthomarina, and Arthrobacter nitroguajacolicus*) were used in microcosm slurry phase bioremediation trials. After 30 days, the degradation rate ranged from 97.8 to 99.4%, where the total TPH concentration was reduced from 30,703 to (170–664) mg kg^{-1}. In addition, the complete biodegradation of the carcinogenic and mutagenic fractions occurred. The research suggests that this sustainable remediation technology can be used as a substitute for the currently used physical and chemical treatment methods applied in Libya (Mansur 2015).

In another study, Mansur et al. (2015) attempt to develop a slurry phase bioremediation protocol. Two hydrocarbonoclastic bacterial isolates, *Pseudomonas* spp. and *Pseudomonas xanthomarina,* were used in three different strategies, namely: *bioaugmentation* (BA), *biostimulation* (BS), and *biostimulation-bioaugmentation* (BS-BA), to assess their ability to reduce the TPHs in COBTS contaminated soil. The results indicated that the BS–BA treatments showed the highest reduction (96–97%, from 30,703 to (860–1020) mg kg^{-1}), followed by the BS treatment (92–93% reduction).

BS–BA treatments are used to assess their ability to reduce the TPHs in COBTS contaminated soil. The results indicated that the BS-BA treatments showed the highest reduction (96–97%, from 30,703 to (860–1020) mg kg^{-1}), followed by the BS treatment (92–93% reduction).

The efficiency of different bioremediation processes was assessed in diesel-contaminated soils. The greatest TPHs degradation, 96.1% and 95%, were obtained when the soil was amended with pea straw alone and in combination with a *hydrocarbonoclastic*, respectively. Meanwhile, natural attenuation resulted in a significantly lower (TPHs) reduction of 76%. The presence of pea straw also led to a significantly increased recovery of hydrocarbon degraders. Metagenomics analysis showed that original soil contained hydrocarbon degraders (e.g. *Pseudoxanthomonas* spp. and *Alcanivorax* spp.). However, they require a biostimulant (in this case pea straw) to become active (Koshlaf et al. 2016).

Microorganisms were also isolated and identified from contaminated soil samples obtained from an oil refinery (Zawia). They were identified as *Pseudomonas putida, Pseudomonas species, Betaproteobacterium, Actinomyces species,* and *Bacillus species*. Among the five species tested, *Pseudomonas putida* showed superior performance in terms of growth on hydrocarbons (1.0×10^{10} CFU/ml), emulsifying activity (86%), and ability to transform hydrocarbons in pure culture. *P. putida* also showed a decrease in heavy hydrocarbon fractions, demonstrating a clear potential for this microbe to be used as a soil inoculant in bioremediation (Althalb and Ian 2017).

7.7 Conclusion

Reviewing these brief summaries from studies and research may give a general idea of Libya's location and climate, as well as the environmental conditions prevailing in the region, which characterize Libya by a biological diversity of different living organisms. In summation, the most important challenges to biological activity in Libyan soil are drought, salinity, and high temperatures prevailing in arid and semi-arid regions that are waiting for more solutions. Through the intensity of studies and research in the three regions (east, west, and south) in the field of soil microbiology, one can assess the extent of interest in this field. What has been presented here is only a small selection of the hundreds of research studies that have been previously and recently published and carried out on soil microbiology. Many of these studies were inaccessible either because they were not published or were published in reviewed journals but unavailable online, in addition to hundreds of masters and doctoral thesis, discussed in the past four decades.

Acknowledgements First of all, we would like to extend our sincere thanks and gratitude to Dr. Hamdi A. Zurqani, who gave us the opportunity to write this part of the book and participate in it through our contribution to collecting as much data and studies as possible that were previously published. We would also like to express deep appreciation to him for his great efforts in following this book patiently and with great interest until the final stage, in addition to his contribution to the linguistic amendments and corrections.We would also like to express our gratitude and deep appreciation to Dr. Mikael Youssef Al Fitouri, a professor at Omar Al-Mukhtar University (Al Bayda—Al Jabal Al Akhdar—Libya) for his cooperation and assistance in obtaining some references and research in the eastern region of Libya. Thanks also go to everyone who supported this work and helped to accomplish it, and to everyone who provided advice or assistance that contributed to its achievement.

References

Abdallah NS, Saeed MA, Elsanousi OM (2005a) Isolation and identification of phython pathogenic fungi accompanied with seeds of some leguminous species. Al-Mukhtar J Sci 12:83–100

Abdallah NS, Saeed MA, Saleh AM (2005b) Chemical control of Fusarium solani isolated from Local–Bean seeds in laboratory and in greenhouse conditions. Al-Mukhtar J Sci 12:101–114

Abdallah NS (2006) Isolation and pathogenicity test for pathogenic fungi carried on some chickpea cultivars growing at Al Jabal Al Akhdar. Al-Mukhtar J Sci 13:90–98

Abdulla NS (2007) Biological control to some seed born fungal pathogens on some legumes. Al-Mukhtar J Sci 14:91–114

Abdelnaby M, Elnesairy NNB, Mohamed SH, Alkhayali YA (2015) Symbiotic and phenotypic characteristics of rhizobia nodulaing cowpea (Vigna Unguiculata L. Walp) grown in arid region of Libya (Fezzan). J Environ Sci Eng 4:227–239

AL-Ani, Mehjin AM (2016) Distribution of the arbuscular mycorrhizal fungi in Al Jabal Al akhdar area, east Libya. Kirkuk University. J Sci Stud (11)3:115–135

Al Hammadi ESM (2014) Identification the impact of growth of different plants on the microbial community in both soil and rhizosphere. Assiut J Agric Sci 45(5):90–102

Al Hammadi ESM (2015) Study the effect of insecticides commonly used locally on the bacteria of economic importance in the soil. Ass Univ Bull Environ Res 18(1):29–42

Al-Idrissi M, Sbeita A, Jebriel A, Zintani, Shreidi A, Ghawawi H, Tazi M (1996) Country report to the FAO international technical conference on plant genetic, Tripoli

Altayyar AI, Osman NA, Elbreki MF, Ibrahim H, Aboalasad A, Barkah A, Almahdi N (2016) Isolation and identification of soil keratinophilic fungi from different area in South of Libya. Int J Appl Med Biol Res 1(1):27–32

Althalb H, Ian S (2017) Isolation of indigenous hydrocarbon transforming bacteria from oil contaminated soils in Libya: selection for use as potential inocula for soil bioremediation. Int J Environ Bioremediation Biodegrad 5(1):8–17

Ashour AA, Mohamed MA, Hami MA (1981) Libyan truffles terfezia boudieri chatin chemical composition and toxicity. J Food Sci 43 (19):927–929

Atallah T, Rizk H, Cherfane A, Firas B, El-Alia R, Lajuide P, Hajj S (2008) Distribution and nodulation of spontaneous legume species in grasslands and shrublands in Mediterranean Lebanon. J Arid Land Res Manag 22(2):17–26

Attitalla IH, Salleh B (2010) Improvement of carboxymethyl cellulase and xylanase production by alginate immobilized Trichoderma harzianum. Biotechnology 9:529–532

Attitalla IH, Abobaker M, Alhasin MA, Nasib AH, Ghazali L, Zakaria HM, Jais I, Balal AA, Salleh B (2010a) Occurrence and microbiological characteristics of Azospirillum strains associated with leguminous and non-leguminous plants in Al Jabal Al Akhdar eco-region, Libya. Am Eurasian J Agric Environ Sci 8(6):617–625

Attitalla IH, Al Ani LKT, Nasib MA, Balal IA (2010b) Screening of fungi associated with commercial grains and animal feeds in Al Bayda Governorate, Libya. World Appl Sci J 9(7):746–756
Attitalla IH (2011) Effects of Azospirillum lipoferum strain isolated from Al Jabal Al Akhdar region and wheat straw on some physiological properties and nitrogen content of Triticum aestivum cultivars. Al-Mukhtar J Sci 26:9–25
Attitalla IH, Abdelrawaf SS, Omar KS, El-Komy HMA, Sarwar M (2012) Occurrence and microbiological characteristics of Trichoderma in Al Jabal Al Akhdar region, Libya. J Biol Sci 12(4):209–217
Alzobaidy DH, Mohammad AM (2003) Agricultural education-actuality and Ambition. Al-Mukhtar J Sci 10:9–24
Baiu H, Nair GA (1999) Isolation of cellulose decomposing bacteria from Libyan soils. Ecol Environ Conserv 5:157–159
Ben Mahmoud MT, Ferjani EA (2016) The effect of bio-fertilization with Azotobacter Spp. and different rates of urea fertilizer on the growth of Barely. Syrian J Agric Res 3(1):213–216
Ben Mahmoud MT (2016) The effect inoculation with bacteria Burkholderia Spp. and different rates of nitrogen fertilization on the productivity of tomato. Syrian J Agric Res 3(2):335–242
Ben Mahmoud MT, Ferjani EA (2017) The effect of different rates of two pesticides (Benomyl and Abamectin) on growth and number of N2 fixation bacteria cells. Syrian J Agric Res 4(2):150–159
Ben Mahmoud MT, Ferjani EA (2018) Influence of soil pH on azotobacter population with using microbiological characteristics as bio-measurement in arable lands of Tripoli. N.W. Libya. Al-Mukhtar J Sci 33:146–154
Ben Mahmoud MT, Ferjani EA, Elalam MF (2018) Inoculation of Grass-cycling compost with N2 fixing bacteria. J Arab Am Univ 4 (1):33–47
Bouzadi M, Grebenc T, Turunen O, Kraigher H, Taib H, Alafai A, Sbissi AME, Bedade D, Shamekh S (2017) Characterization of natural habitats and diversity of Libyan desert truffles. Biotech 7 (5):218–328
Buhidma MS, Abogharsa IA, Saeed MA (2007) Biological and chemical control of Fusarium wilt fungus on tomato cultivars. Al-Mukhtar J Sci 16:151–162
El-Buni AM, Rattan SS (1981) Check list of Libyan fungi, Department of Botany Faculty of Science, University of Al Faateh, Tripoli, Libya
El-Giahmi AA, Nicolson TH, Daft MJ (1976) Endomycorrhizal fungi from Libyan soils. Trans Br Mycolog Soc 67(1):164–169
El Sadi SMA (2004) Characteristics of root-nodulation bacteria isolated from Acacia karroo in Fezzan region of Libya. MSc Thesis, Faculty of Sciences, Sebha University, Sebha Libya
El-Said AHM, Saleem A (2008) Ecological and physiological studies on soil fungi at western region, Libya. Mycobiology 36(1):1–9
Elssaidi MA, Mohamed HY (2015) Effect of Some pesticides on nitrogen fixing bacteria under Wadi Alshatti Conditions, Libya. Marine Sci J Environ Technol 1(2):70–85
Environment Public Authority (2008) National strategy for sustainable development, January (2008)
Faituri MY, El-Mahi YE, El-Hassan GA (2001) Effects of some salts and sodicity on the growth of a Rhizobium leguminosarum bv. viceae strain isolated from a salt-affected soil. Canad J Microbiol 47 (90):807–812
Fiatori MY, El-Hassan GA, El-Mahi YF (1996) Effect of salinity and sodicity on bacterial-plant symbiotic relationship. Al-Mukhtar J Sci 3:112–128
Hesnawi RM, Mogadami FS (2013) Bioremediation of Libyan crude oil-contaminated soil under mesophilic and thermophilic conditions. APCBEE Procedia 5:82–87
Ibrahim FA, Hussin AI, Al Amin SK (2019) Study of physiological characteristics of rhizobia isolated from Melilotus officinalis L. growing in arid soil of Libya (Fezzan). A thesis Submitted in Partial Fulfillment of the Requirements of Master Department of Botany Faculty of Science Sebha University Sebha, Libya
Imam MK, Badawy FH (1978) Response of three potato cultivars to inoculation with Azotobacter. Potato Res 21(1):1–8
Jafri SMH (1978) Mimosaceae. In: Jafri SMH, El-gadi A (eds) Flora of Libya, vol. 60. Faculty of Science Al fateh University Tripoli, Libya
Jadallah SO, Abdelall MF, Elkobeisy OS, Ali MA (2017) Phenotypic and phylogenetic diversity of mycorrhizae isolated from Al Jabal Al Akhdar, Libya. Biosci Res 14(4):915–923
Khalifa MOA (2013) Phenotypic diversity of rhizobia nodulating some wild leguminous plants growing in Libya. A thesis Submitted in Partial Fulfillment of the Requirements of the degree of Master. Department of Botany Faculty of Science University of Sebha. Sebha, Libya
Khalifa MO, Jarma A, Nseer AE, Aboslh MM, Saed NA, Asheiw JI, Ahmadi IM (2018) The Isolated root nodulating bacteria from Medicago sativa L. and its production of enzymes. JOPAS 17 (1):275–281
Koshlaf E, Shahsavari E, Medina AA, Taha M, Haleyur N, Makadia TH, Morrison PD, Ball AS (2016) Bioremediation potential of diesel-contaminated Libyan soil. Ecotoxicol Environ Saf 133:297–305
Kran J (1965) A list of plant pathogenic and other fungi of Cyrenaica (Libya) Phytopathological Papers
Makawi AAM (1973) The density of azotobacter in root-free soils and in the rhizosphere of several plants in arid and semi-arid Libyan regions. Allgemeine, Landwirtschaftliche und Technische Mikrobiologie 128(1–2):135–139
Makawi AAM, Abdel-Nasser M, Abdel-Moneim AA (1979) Quantitative effects of some pesticides on certain physiological groups of micro-organisms in soil. Mikrobiologie der Landwirtschaft, der Technologie und des Umweltschutzes 134(3):223–228
Mansour AMA (2010) Contribution to knowledge of some soil fungi in eastern region in Lybia. J Product Dev 15(3):395–404
Mansur A (2015) Bioremediation of Libyan soil contaminated with crude oil tank bottom sludge. PhD Thesis, Applied Science, RMIT University, Australia
Mansur AA, Adetutu EM, Makadia T, Morrison PD, Ball AS (2015) Assessment of the hydrocarbon degrading abilities of three Bioaugmentation agents for the bioremediation of crude oil tank bottom sludge Contaminated Libyan soil. Int J Environ Bioremed Biodegrad 3(1):1–9
Marei A (2019) Indignous Mycorrhizal spores isolation and identification from wheat field, Benghazi District Libya. Int J Agric Environ Res 5(3):340–346
Mlitan AB (2013) Heavy metals and microbial toxicity of analysis of soil samples near iron and still factory (Misurata, Libya). In: Second International Conference on Chemical, Environmental and Biological Sciences (ICCEBS'2013) March 17–18, 2013. Dubai (UAE)
Mlitan AB, Alajtal AI, Alsadawy AM (2013) Toxicity of heavy metals and microbial analysis of soil samples collected from the area around Zliten cement factory. Open J Air Pollut 2(1):25–28
Mlitan AB, Abofalga A, Swalem A (2015) Impact of treated wastewater on some physicochemical parameters soil and its fungal content. Int J Environ Sci Dev 6(5):369–380
Mohamed SH, Smouni A, Neyra M, Kharchaf DA, Maltouf F (2000) Phenotypic characteristics of root-nodulating bacteria isolated from Acacia spp. grown in Libya. Plant Soil 224:171–183
Mohamed OA (2018) Effect of root-nodule bacteria on growth of acacia saligna under salt stress. Al-Mukhtar J Sci 33(3):250–256
Mohamed SH, Bertata NZ, Khalifa MO (2018a) Selection of salinity tolerant rhizobial isolates from wild and cultivated legumes in arid region of Libya (Fezzan). J Pure Appl Sci JOPAS 17(1):308–317

Mohamed SH, Khalifa MOA, Babiker NN (2018b) Physiological and symbiotic characteristics of rhizobia isolated from medicago polymorpha L and trigonella stellate L growing in semi-arid regions of Libya. Adv Biosci Bioeng 6(4):34–41
Moubasher AH (2010) Endangered desert truffles in Egypt and neighbouring Arab countries, with further notes on their distribution. Mycol Balc 7(1):59–64
Mouchacca J (2005) Mycobiota of the arid middle east: check-list of novel fungal taxa introduced from 1940 to 2000 and major recent biodiversity titles. J Arid Environ 60(3):359–387
Naim MS (1965) Development of rhizosphere and rhizoplane microflora of Aristida coerulescens in the Libyan Desert. Archiv für Mikrobiologie 50(4):321–325
Naim MS (1967) Contribution to the knowledge of soil fungi in Libya. Mycopathologia et mycologia applicata 31(4):296–299
Naim M S (1976a). Contribution to the knowledge of soil fungi in Libya. I. Rhizosphere and soil fungi of Artemisia herbaalba in Tripoli. Mycopath. Mycol. Appl. (31):296–299
Naim M S (1976b) Contribution to the knowledge of soil fungi in Libya. II. Fungus flora under Citrus trees in Libya. Mycopath Mycol Appl (31):296–299
Sedra MH (2010) Evaluation of soil receptivity of date palm groves in Arab countries to Fusarium oxysporum f. sp. albedinis, causal agent of Bayoud disease of date palm. In: IV international date palm conference, vol 882, pp 515–525
Selim, Nour El Dein MS (1971) Search for antagonistic Actinomycetae in Libyan soils. Canad J Microbiol 17(5):731–735
Selim, Nour El Dein MS, Khalil GA (1979) Investigations into rhizosphere microflora of some plants in Libya. Mikrobiologie der Landwirtschaft, der Technologie und des Umweltschutzes 134 (3):237–242
Shaban G M (1986) Physiological and ecological studies on the genus Trichoderma in Egypt soils. PhD Thesis, Faculty of Science, El-Minia University, El-Minia, Egypt
Shaieb FM, Elghazawani AH, Issa A (2015) Studies on crude oil degrading bacteria isolated from Libyan desert. Int J Current Microbiol Appl Sci 4(2):920–927
Shoaib FM, Fitouri MY, Attiat Allah EH (2009) Tolerance of some Rhizobium leguminosae bv. Vicia strains to different salt concentrations. Al-Mukhtar J Sci 23:127–140
Taher AA, Abubaker NS, Hesham MA, El-Komy (2019) Screening of different isolates of Azospirillum for growth at different salt (Nacl) concentrations. The third international conference on basic science and their applications. Special conference issue, vol 1, no 1, pp 454–462
Watanabe T, Uematsu S, Inoue Y (1988) Pathogenicity of twenty-three Pythium isolates from soils of Shikoku Island. Trans Mycol Soc Jpn 54:565–570
Watanabe T (2002) Pictorial atlas of soil and seed fungi, Morphologies of cultured fungi and key to species, 2nd edn. CRC Press
Youssef YA (1974) On the fungal flora of Libyan soils. Arch Microbiol 99:167–171
Zahran HH (1991) Conditions for Successful Rhizobium-legume Symbiosis in saline environments. Biol Fertil Soil 12:73–80
Zahran HH (1999) Rhizobium- legume symbiosis and nitrogen fixation under severe condition and in arid climate. Microbiol Mol Biol Rew 63:968–989
Zaid AM, Mayouf M, Said YF (2014) The effect of pre-emergent herbicides on soil microflora and N-fixing bacteria in pea field. Int J Sci Basic Appl Res 15:131–138
Zekry Y (2005) Libyan Climate an applied study of physiological climate patterns. PhD Dissertation, Department of Urban development Faculty of Earth Sciences, Geography, University of Mentouri. Constantine, Algeria
Zurqani H, Mikhailova E, Post C, Schlautman M, Sharp J (2018) Predicting the classes and distribution of salt-affected soils in north-west Libya. Commun Soil Sci Plant Anal 49(6):689–700
Zurqani HA, Mikhailova EA, Post CJ, Schlautman MA, Elhawej AR (2019) A review of Libyan soil databases for use within an ecosystem services framework. Land 8(5):82

Dr. Eman Ali Ferjani is an associate professor at the Department of Soil and Water Sciences, Faculty of Agriculture, University of Tripoli, Tripoli, Libya. She received the M.Sc. (Agricultural Science) in 1997 from University of Tripoli, Tripoli, Libya. Dr. Ferjani completed her Ph.D. (in Microbiology) from science and biotechnology, School of Microbiology, UKM University, Kuala Lumpur, Malaysia in 2011. She started her career as a researcher in environmental microbiology since 1990. She received intensive training from The International Center for Agricultural Research in the Dry Area (ICARDA) in the field of biofertilizer and biological nitrogen fixation in 1989/1990. Dr. Ferjani held the position of head of the Basic Sciences Department of the Faculty of Agriculture, University of Tripoli, Tripoli, Libya from 2013 to 2018. Her major research and teaching activity focused mainly on soil microbiology, biotechnology and soil, water, and plant analysis in addition to soil fertility. Her current research centers on environmental pollution mainly on the biodegradation of waste oil, plastics, and heavy metals in soil. She teaches courses of microbiology, soil microbiology, and soil, water, and plant analysis, and has supervised the graduation of master's and bachelor's degree students. Dr. Ferjani is also working as an associate professor in the Department of Environmental Engineering, Libyan Academy, Tripoli.

Dr. Merfat Taher Ben Mahmoud is an assistant professor of soil science at the Department of Soil and Water Sciences, Faculty of Agriculture, University of Tripoli, Tripoli, Libya. She received her M.Sc. (Agricultural Sciences) in 1998 from University of Tripoli, Tripoli, Libya. Dr. Ben Mahmoud completed her Ph.D. in (Soil Microbiology) from RMIT University, Melbourne, Australia, in 2009. She specialized in biotechnology and environmental biology. Her main research interests focused on soil fertility, plant nutrition, soil microbiology, and biofertilizer studies. She has been teaching courses and has supervised the graduation of master's and bachelor's degree students since 2009.

Dr. Asma Alnajjar, Ph.D. in Molecular Biology, is an associate professor at the Department of Soil and Water Sciences, Faculty of Agriculture, University of Tripoli, Tripoli, Libya. She has over 30 years of experience as a researcher in environmental microbiology and biotechnology. Her past research subjects were on harmful plant–microbe interaction, antimicrobial effects of native medicinal plant extracts, biodegradation, and decolorization of synthetic dyes. Her current research is focusing on improving soil quality through the study of the role of soil microorganisms in nutrition and protection, the effect of agrochemicals on non-targeted soil microorganisms, and soil biocementation. Future research will concentrate on regenerative agriculture in arid and semi-arid regions for food security. She teaches graduate and undergraduate courses of microbiology, biochemistry, molecular biology, and biotechnology. Apart from her research and teaching activities, she has participated as one of the team for the establishment of the biotechnology research center in Tripoli. For the past few years, she has served as a department graduate coordinator and also director of the graduate and training office in the faculty of agriculture.

Land Degradation and Desertification

8

Basher A. Nwer, Hamdi A. Zurqani, and Azalarib S. Ali

Abstract

Desertification and land degradation represent a major threat to agricultural sustainable development. It puts the livelihoods of billions of people around the world at risk in arid and semi-arid regions. This is especially the case for people living in rural areas where the majority of the poor live in rural areas. Human pressures and climate variations are the main causes of land degradation and desertification. It occurs because drylands ecosystems are extremely vulnerable to over-exploitation and inappropriate land use that resulted in underdevelopment of economies and in entranced poverty among the affected population. Libya is no different from other parts of arid and semi-arid zones in terms of the climatic conditions and human activities as the main reasons behind land degradation and desertification. However, many human factors are contributing to the deterioration of environmental conditions and the occurrence of desertification in Libya. This chapter discusses land degradation and desertification in Libya, emphasizing the main causes, impacts of the phenomena, and efforts to combat it.

Keywords

Libya • Mediterranean region • Population • Sahara desert • Soil resources • Soil erosions • Salinization

B. A. Nwer · H. A. Zurqani (✉) · A. S. Ali
Faculty of Agriculture, Department of Soil and Water Sciences, University of Tripoli, Tripoli, 13538, Libya
e-mail: h.zurqani@uot.edu.ly

B. A. Nwer
e-mail: bashir.nwer@yahoo.com

A. S. Ali
e-mail: az_alarib@yahoo.com

H. A. Zurqani (ed.), *The Soils of Libya*, World Soils Book Series,
https://doi.org/10.1007/978-3-030-66368-1_8

8.1 Introduction

Land degradation and desertification are among the most serious global, regional, and local environmental issues (UNEP 1997). Nowadays, desertification is considered to be one of the greatest environmental challenges. It constitutes a major difficulty in meeting basic human needs and sustainable agriculture development in semi-arid and arid lands. It has severe consequences on many living aspects such as food security, economic activity, infrastructure, human health, natural resources, and the environment. The direct effect of land degradation varies from the reduction in land productivity or the complete abandonment of agricultural land. The desertification leads to agricultural productivity losses and increases poverty. This could cause a food crisis facing wide areas in arid regions.

Land degradation puts the livelihoods of billions of people around the world at risk (IPBES 2018). Over 250 million people are directly affected by desertification and one billion people in over 100 countries are at risk. This is especially the case for people living in rural areas where the majority of them are poor. Estimates report that 80% of the extremely poor live in rural areas and 65% work in the agricultural sector (Castaneda et al. 2018).

Desertification is recognized as a complex phenomenon that requires researchers' expertise in disciplines such as climate science, soil science, meteorology, hydrology, range science, agronomy, veterinary medicine, geography, political science, and economics. Researchers in these and other disciplines, as well as many national disciplines, have defined it in many different ways. As a result, there were a number of definitions for desertification; for instance, at the 1992 Rio Conference it was defined as degradation of sensitive drylands resulting from several factors, including climate variations and human activities. The United Nations Convention to Combat Desertification (UNCCD) has defined desertification as "land degradation in arid, semi-arid and dry sub-humid areas resulting from various factors, including climatic variations and human activities" (UNCCD, Article 1, 1994). All these definitions agree that desertification can be caused by the complexity of climate variations and human activities. MEA (2005) stated that desertification is caused by a combination of factors, which change over time and vary by location.

Desertification occurs in drylands around the world and its impacts are felt locally, nationally, regionally, and globally. Drylands occupy 41% of Earth's land area where more than 2 billion people live. This represents a third of the human population in the year 2000. The drylands include all terrestrial regions where water scarcity limits the production of crops, forage, wood, and other ecosystem provisioning services.

Human pressures on the environment and climate crises are the main reasons for the degradation of natural resources. These crises lead to social changes, modified consumption habits, the search for non-farming work, and temporary labor migration, as well as the sale of productive assets and permanent migration. Societies affected by desertification suffer declining incomes and decreasing food security (UNCCD 2007). Drought and desertification are at the core of serious challenges and threats facing sustainable development in Africa. These problems have far-reaching adverse impacts on human health, food security, economic activity, physical infrastructure, natural resources and the environment, and national and global security.

The African continent is the most affected by land degradation and desertification. Reich et al. (2001) stated that desertification processes affect about 46% of Africa. This number is very important, taking into consideration that about 43% of the continent is characterized as extreme deserts. Only about 11% of the land can be excluded from desertification processes. Rising temperatures are forecasted by IPCC in North Africa which could have a severe impact on the regions including Libya. This could have an impact on the regions already facing stressed resources including water and food (IPCC 2007). In addition to climate change, misuse of water and soil resources, urbanization, overgrazing, removal of natural vegetation, and rapid changes in land use are important reasons for increasing land degradation and desertification in Libya. This chapter discusses the causes and effects of land degradation and desertification and effort to combat desertification and land degradation in Libya.

8.2 Desertification Land Degradation and Status in Libya

According to The National Plan to Combat Desertification (ACSAD and Libyan experts 2005), much of Libya's agricultural land is under pressure from either soil erosion, loss of natural vegetation cover, or over-use of irrigation water. Problems such as soil salinity and rising groundwater all appear to be on the rise. The environmental impacts of agricultural activity are the result of a complex chain of biophysical and other factors, which are linked to the natural characteristics of the land. The total land degradation areas from 1981 to 2006 were about 1,024,053 km^2 (Table 8.1).

Libya is suffering from desertification in various types and degrees. It can be noted that many areas of Libya were exposed to the over-use of the natural resources which led to their deterioration and the acceleration of desertification problems in these areas. The types and aspects of desertification in Libya are as follows (Ben Mahmoud 2013).

8.2.1 Soil Erosion by Wind

It is considered to be the most common environmental problem in Libya with respect to the loss of the fertile surface layer or the problems caused by the transfer, encroachment, and fall of the sands, especially for the population centers in addition to their adverse effects on public health.

8.2.2 Soil Erosion by Water

This phenomenon leads to the loss of the fertile surface layer and the transfer of large quantities of soils to other sites leaving rocks, shallow soils, or bare lands. The eroded materials settle in certain areas behind the dams or in the agricultural fields, causing destruction and forming another deterioration aspect.

8.2.3 Loss of Nutrients

The agricultural intensification and failure to abide by the agricultural instructions lead to the decline of soil fertility and the loss of vast areas of the agricultural lands. The loss

Table 8.1 Statistics of degrading areas in Libya (1981–2006) based on (GTZ 2009)

Country	Area in (km^2)	% Total population	LD areas (km^2)	Improved areas (km^2)	Affected people
Libya	1.759.540	6.310.434	1024052.28	527.86	14651730

of nutrients in the irrigated areas leads to the decrease of the production capability of the lands and their deterioration at varying degrees.

8.2.4 Soil and Water Pollution

Soil and water resources in Libya are exposed to many pollutants that differ according to the types of the prevailing farming systems and the techniques used by farmers. This problem becomes clearer when fertilizers and pesticides are used irrationally, especially in the areas of the irrigated agriculture and where the industrial constructions are close to agricultural fields, as this leads to the settlement of the pollutants resulting from these constructions either on the soil surface or in the soil.

8.2.5 Salinization

Salinization is also the primary type of soil degradation in Libya. This problem becomes clear in the irrigated areas or when the water table is increased. It starts with the accumulation of salts which make the soils not suitable for growing crops, and after sometime these soils become desertified.

8.2.6 Aridification

The depletion of the groundwater resources, especially those near the surface, leads to the aggravation of the desertification and aridity processes and adversely affects the quantity of the available waters. When these waters become dry or of bad quality, the people go elsewhere and leave the lands exposed to desertification and aridity.

8.2.7 Deforestation

The forested and shrublands areas have been suffering from severe deforestation, and that is considered one of the biggest environmental challenges in the country (Zurqani et al. 2019). Deforestation increases soil erosion rates due to exposure to soil mineral by removing the humus and litter layers from the soil surface.

8.3 Causes and Consequences of Desertification

8.3.1 Causes of Desertification

Literature has widely discussed land degradation's causes (Thomas 1997; Lambin et al. 2001; Reynolds and Stafford Smith 2002; Geist and Lambin 2004; Ben Mahmoud and Lolo 2010). It has been concluded that desertification is caused by multiple direct and indirect factors. It occurs because drylands' ecosystems are extremely vulnerable to over-exploitation and inappropriate land use, which results in underdevelopment of economies and in entranced poverty among the affected population. Whereas over-cultivation, inappropriate agricultural practices, overgrazing, and deforestation have been previously identified as the major causes of land degradation and desertification, it is, in fact, a result of much deeper underlying forces of socio-economic nature, such as poverty and total dependency on natural resources for survival by the poor. It is also true to state that desertification problems are best understood within the dictates of disparities of income and access to or ownership of resources. The primary cause of deforestation and natural pastures in the country is due to the rapid urbanization process and wood extraction for charcoal (Fig. 8.1a, b).

According to the Global Forest Watch, the total area covered by forest trees was about 24,344 hectares in 2000, and it decreased over time as it was about 12,107 hectares in 2011 and 11,866 hectares in 2018. Figure 8.2 shows deforestation activities in the north-western coastal areas stretching from Tajoura to Misurata (north-east Tripoli): (a) before December 2006, (b) after December 2016.

In general, the environment in Libya is vulnerable. The variability of rainfall and the occurrence of occasional relatively heavy showers characterized as high intensity can produce runoff. The removal of natural vegetation from the land surface is the main factor that accelerates soil erosion. The combination of these factors in addition to the topography has increased the rate of soil erosion by water in this area (Nwer 2005).

Subsequently, the causes of desertification are too complex to be simplified. Desertification is driven by a group of main variables, mainly climatic factors (Yang and Prince 2000; Hulme and Kelly 1993). The natural factors include changes in climate: mainly rain, wind, and temperature.

Fig. 8.1 Deforestation activates; **a** cutting trees for charcoal production, and **b** charcoal packed in white polypropylene bags

These factors vary from one site to another in response to changes in external and internal influences in the air system (Emgaili 1993). For example, in the spring and autumn, strong southerly winds, known locally as “Ghibli” blow from the desert, filling the air with sand and dust and raising the temperature to approximately 50 °C (Fig. 8.3). These strong winds are a major erosion factor in the desert, transporting sand from one place to another.

As previously mentioned in the climate section, the climate in Libya can be divided into two main climate conditions: arid climate inland and Mediterranean climate on the coast, see Chaps. 1 and 3. As for aridic conditions and drought, it is well documented and discussed. However, the nature of the Mediterranean climatic condition could contribute to land degradation and desertification. According to Sevink (1988), climatic characteristics of the Mediterranean region include rare freezing, hot summers with at least two to three dry months, and cool rainy winters; precipitation often falls as storms of high intensity which produce torrential runoff (Bradbury 1981). Because of these violent storms, the Mediterranean climate is described as one of the most aggressive with respect to erosion. Also, in regions

Fig. 8.2 Deforestation in the north-western coastal areas stretching from Tajoura to Misurata (north-east Tripoli): **a** before December 2006, **b** after December 2016. Google Earth imagery (http://www.earth.google.com)

(a)

(b)

such as the southern Mediterranean, cracks can form by desiccation during dry summers, causing extreme dissection of the slopes. A major problem in the climate in this region is that the winter rainfall, which causes erosion, does not coincide with the vegetation cover that protects the soil surface, especially in cultivated cropland and heavily grazed pasture. The Mediterranean climates do not favor the development of a dense vegetation cover on most slopes, which are poorly stabilized at ground level. As a result, areas with Mediterranean type climates are traditionally classified as areas with high potential erosion rates (Saunders and Young 1983; Brown 1990).

Libya is no different from other parts of arid and semi-arid zones in terms of the climatic conditions and human activities being the main reasons behind land degradation and desertification (Zurqani et al. 2019). However, many human factors are combined, contributing to the deterioration of environmental conditions and the occurrence of desertification in Libya. Urbanization can intensify land degradation and desertification. FAO (2015) stated that urbanization has led to over 25% of highly fertile lands being converted to urban areas.

The discovery of oil and development of the agriculture sector in the country has also contributed to the process of

Fig. 8.3 Wind with sand and dust, known locally as "Ghibli"

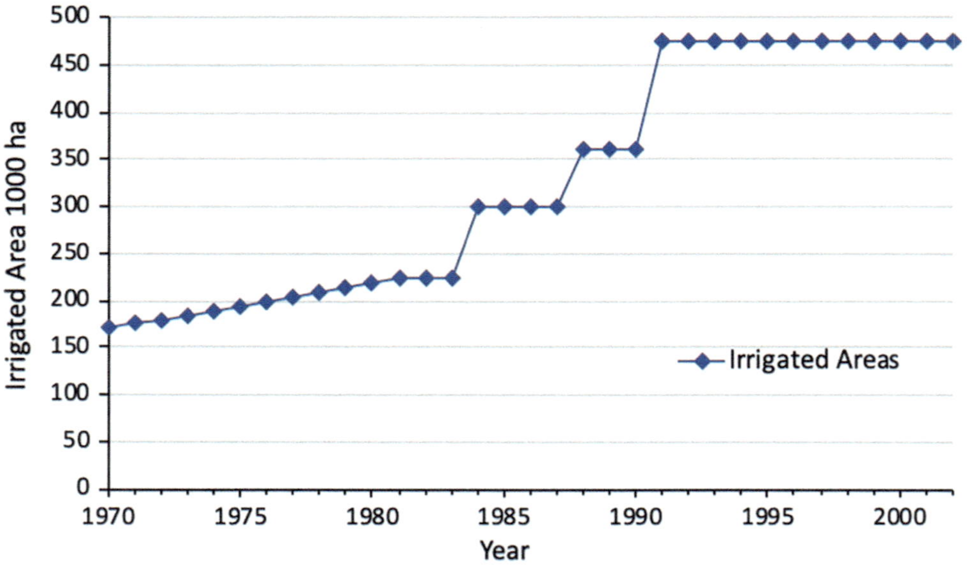

Fig. 8.4 Irrigated Areas in Libya from 1970 to 2002 (adapted from Nwer 2005)

land degradation and desertification. The wealth created by oil has led to a huge investment in agriculture. Nevertheless, agricultural development has put severe pressure on local ecosystems, especially in sensitive and fragile areas. This was because of the change in production patterns, and the need to produce more food for the growing population. The rapid expansion of the irrigated areas in Libya has put pressure on resources already at stress. The irrigated area has increased four times in 2002 compared with 1970. Figure 8.4 shows the rapid increase in the irrigated areas from 1970 to 2002.

It can be seen that the irrigated area was less than 160,000 hectares in the 1970s, whereas in 1986 it has increased to 300,000 hectares. Compared with 1970 and 1986, the irrigated area in 2002 has rapidly increased three times and four-times, respectively. It has expanded to more than 450,000 hectares. Consequently, this led to the intensification of pressure on already limited natural resources, thus escalating land degradation and desertification problems (Ben Mahmoud et al. 2003).

This section has explained the causes of land degradation and desertification in general with specific reference to nature and caused in Libya. Land degradation and desertification in causes are climatic variations and human activities. However, many human factors such as increased human population, urbanization and intensive agriculture, are combined contributing to the deterioration of environmental conditions and the occurrence of desertification in Libya.

8.3.2 Consequences of Desertification

The impact of the land degrading process varies according to the natural characteristics of the land. These characteristics include soil type, slope, vegetation, and climate. Therefore, the degradation process could take place in one site and may not cause the same degradation in another site. This is due to the fact that soil characteristics, topography, and climatic conditions are different. So, similar rates of soil loss will be different as a result of erosive rainstorms occurring above different soil types and slope degrees. Therefore, identification of reasons behind land degradation must recognize the natural interactions between different elements in the landscape which affect degradation, and also the site-specificity of degradation. For instance, thunderstorms and flash floods increase the problem of soil erosion by water extensively (IPCC 2001; GTZ 2009).

Land degradation adverse consequences are both environmental and socio-economic. It weakens the structure and functions of environmental systems such as the biogeochemical cycles (i.e. carbon, hydrological, and nutrient cycles) which are threatening the survival of human beings. This put at risk more than 1 billion people in developing countries in their livelihoods and economic wellbeing, and their nutritional status (World Bank 1998). According to ACSAD (2007), more than 40% of the Libyan population are affected by desertification.

According to the UN (1992), land degradation can also cause food and water scarcity, loss of income, resource conflicts, and environmental deterioration. Poverty is closely linked to land degradation. Most of the affected areas are rural poor, and the land is the main provider for their income to survive. Usually, competition among people in those areas takes place for declining natural resources. Therefore, the land becomes further degraded and the cycle of poverty is perpetuated. In Libya, this has caused internal migration in the past from rural areas to urban areas where opportunities for employment and better living conditions exist. As can be seen (Fig. 8.5) in 1961, the urban population in Libya was 28.5% of the total population, whereas in 2017 it increased to 79.8% (FAO STAT 2019). This led to increased pressure on the services of the main cities such as Tripoli and Benghazi, as well as increased the pressure on the groundwater resources in the coastal strip (El-Tantawi 2005).

In addition, one of the most significant causes of the current civil war is fighting for resources. This can be concealed by different political and social reasons. In fact, land degradation and desertification led to poverty and competition on natural resources (oil in Libya). The feeling of injustice in underdeveloped areas ignites conflicts.

Because the causes and effects of land degradation are unaffected by the boundaries of land ownership or use rights, degradation may occur on a farmer's land as a result of actions taken by other land users upslope. Similarly, actions taken on a farmer's field may affect other land users' downslope. The interest in preventing land degradation may not coincide with the cause. Therefore, this will have serious implications when it comes to the costs and benefits of different courses of action are assessed.

The impact of land degradation and desertification is witnessed in Libya in many aspects. Land degradation causes the deterioration of the pastoral environment and a decline in the productivity of livestock, and thus reduces the productivity of meat and milk. Moreover, the deterioration of water resources, soil erosion, and the loss of biodiversity are the main impacts of land degradation. The reduction in biodiversity may contribute to the destruction of the habitats of animal and plant species and micro-organisms.

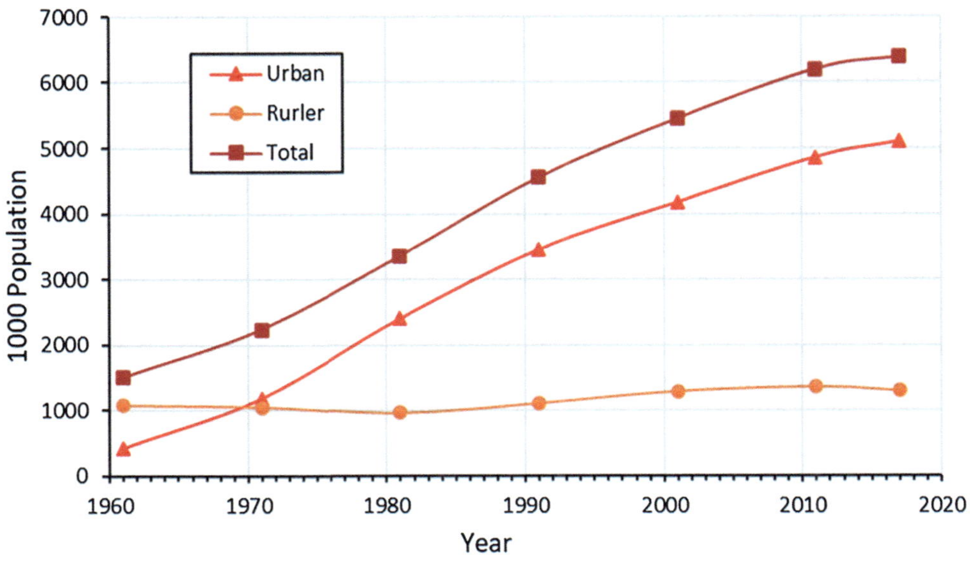

Fig. 8.5 Population growth in Libya (*Source* FAO STAT 2019)

Biodiversity loss is also expected to encourage the genetic erosion of local livestock and plant varieties and species living in fragile ecosystems (Abahussain et al. 2002; Mansour et al. 2011).

The consequences of desertification are deeply felt in rural areas. They can be listed as follows: the decline of soil productivity, intensive erosion, salinization, loss of biodiversity, climate change, and degradation of rangeland (Zurqani et al. 2018, 2019). This will lead to complicated social-economic consequences which have an effect on the stability of the country recently and in the future.

8.4 Efforts to Combat Desertification in Libya

Since the early 1960s, Libya has taken serious measures to combat desertification as part of a broad policy in the framework of the National Plan for Agricultural Development which takes into account the objectives of local development on one hand and the harsh environmental conditions prevailing in the countries on the other. Libya has adopted a number of measures and actions to control desertification during the last four decades by implementing a range of diverse projects in many areas (i.e. forest, pastures, sand dune fixation, soil and water conservation, resistance to erosion and integrated agricultural development).

Libya has conducted a number of measures and projects to control desertification, which include sand dunes fixation, the establishment of windbreaks, reforestation of fallow forest land, establishment of terraces to combat soil erosion, preservation of rainwater on sloping agricultural land, and following of the special agricultural cycle to maintain soil fertility, especially in the areas of cultivation of grain, as well as the protection and improvement of natural pastures (ACSAD and Libyan experts 2005).

Sand dunes stabilization projects have been conducted in Libya since the 1950s. A number of methods and techniques were used to protect lands from moving sands. One of the most famous methods is sand stabilization with oil. This method was used in 1953 when the Forestry Department of Agriculture in Libya began a long-range program of afforestation and sand dune stabilization to reclaim some of the wastelands of the Libyan desert. The basic aim of the program is the creation of a "green belt" 50 km wide extending the entire length of the Libyan coastline from Tunisia to Egypt. In 1953 the Forestry Department reclaimed 60,000 hectares of dune land. Most of the area has been planted with some 36 million acacia and eucalyptus seedlings grown in Libya. In 1960, the Forestry Department working with Esso Standard Libya Inc. initiated an experiment with an oil product which sprayed oil on sand dunes to form a coating that stabilized the dunes for a year. The seedlings were planted before spraying oil. The plant trees then grew to stabilize the sand dunes. The method that was used proved to be successful. This method replaced the traditional method called "dissing". Dissing involves planting seedlings in plots during the rainy winter season and surrounding the plots with windbreaks of dried grass. This method was expensive, slow, and limited (Commonwealth Forestry Association 1969). Libya has embarked yearly plantation campaigns since 1990 which aimed at severely degraded lands in order to rehabilitee lands and establishment of national parks and wildlife to combat desertification and preserve biodiversity.

Other measures that have been taken by Libya were to establish institutions working in combating desertification to plan and organize these efforts. The main task was assigned to the Agriculture Ministry to liaise and coordinate these efforts. This led to the signature of United Nations Conversions to Combat Desertification (UNCCD) in 1984 and to the formation of the National Committee to Combat Desertification (NCCD). NCCD is assigned as the national coordinator to combat desertification and is a focal point to UNCCD.

NCCD is a cross-sector committee formed by members from all interested parties in combating desertification. The NCCD has produced a National Action Plan (NAP) to combat desertification which includes a number of policies, plans, and projects. The NAP is an integrated plan as a part of the development plan of Libya and provides a scientific framework for combating desertification and the development of natural resources within a comprehensive concept that conforms to the Agenda of the twenty-first century and is in line with international efforts to combat and reduce desertification.

Therefore, there was a necessity to develop and update laws and legislative basis to aid efforts to compact desertification and achieve goals and requirements of sustainable development. There are two types of legislation which aid in combating desertification and protecting natural resources. They are natural resource protection legislation and legislation regulating institutions involved in the protection of natural resources. Table 8.2 listed the legislation mentioned.

Despite a dedicated effort from state governments and communities and the range of policy initiatives to promote sustainable natural resource use, Libya still has some significant challenges ahead to achieve ecologically sustainable land management.

Table 8.2 Legislation in Libya (Source NAP 2002)

No.	Legislation	Laws	Objective
1	Natural resource protection legislation	Law (27) 1966	Plant protection
		Law (46) 1975	Protection of small lands
		Law (5) 1982	Protection of Forest and Rangeland
		Law (7) 1982	Protection of Environment
		Law (790) 1982	Organization of Drilling and Water Resources
		Law (1) 1983	Agriculture Technology
		Law (15) 1984	Protection of Animals and Trees
		Law (15) 1992	Protection of Agriculture Land
2	Regulation of Institutions	Law (109) 1971	Establishment of Agriculture Research Centre
		Law (827) 1992	National Research Organization
		Law (72) 1982	Establishment of the Arab Center for Desert Research and development of desert communities.

8.5 Conclusion

Desertification and land degradation is a global phenomenon that is threatening the livelihoods of billions of people around the globe. The causes of land degradation and desertification are well documented and explained. Climate conditions and human activities are the main reasons for desertification. However, the causes of land degradation are too complicated to be simplified. The impact of desertification has countless negative consequences on natural resources. These consequences are likely to become much more complicated unless major new efforts characterized by profound changes in local and international behavior are made. These efforts have to address the livelihood needs of dryland populations and the reversal of the desertification process on war footings through short-term goals and long-term initiatives with a political will and commitment.

References

Abahussain AA, Abdu AS, Al-Zubari WK, El-Deen NA, Abdul-Raheem M (2002) Desertification in the Arab Region: analysis of current status and trends. J Arid Environ 51:521–545

ACSAD and Libyan experts (2005) The national plan to combat desertification (in Arabic)

ACSAD (2007) Desertification in the Arab World. In: Background paper, regional conference on land degradation issues in the Arab region. ESCWA/LAS/UNEP/ROWA/ACSAD (Arab Centre for the Studies of Arid Zones and Dry Lands), Cairo, Egypt 2007 (in Arabic)

Ben Mahmoud KR (2013) Towards a national strategy for the sustainability of natural resources and enhancing food security in Libya, pp 123 (in Arabic)

Ben Mahmoud KR, Lolo A (2010) Desertification and land degradation issues in the Arab region. Arab Centre for the Studies of Arid Zones and Dry Lands), Damascus, Syria

Ben Mahmoud KR, Mansur S, AL-Gomati A (2003) Land degradation and desertification in Libya. In: Alsharhan AS, Wood WW, Goudie AS, Fowler A, Abdellatif EM (eds) Third millennium. Swets and Zeitlinger Publishers. Lisse, The Netherland, pp 339–350

Bradbury DE (1981) The physical geography of the Mediterranean lands. In: DiCastri F, Goodall DW and Specht, (ed) Mediterranean type shrublands. Amsterdam, Elsevier, pp 51–62

Brown CA (1990) Soil erosion and fire in area of Mediterranean type vegetation: results from Chapparral in southern California, USA and Matoral in Andalusia, southern Spain. In: Thrones JB (ed) Vegetation and erosion, process and environments. Wiley, Chichester, pp 269–287

Castaneda A, Doan D, Newhouse D, Nguyen MC, Uematsu H, Azevedo JP (2018) A new profile of the global poor. World Dev

Commonwealth Forestry Association (1969) Sand dune stabilization and afforestation in Libya. Commonw For Rev 48(4) (138):377–381 (December 1969)

El-Tantawi AM (2005) Climate change in Libya and desertification of Jifara plain: using geographical information system and remote sensing techniques. Dissertation, The Johannes Gutenberg University, Mainz, German

Emgaili E (1993) Climate changes and its impact on human and the environment. 1st issue, the University Magazine

FAO STAT (2019). http://faostat.fao.org/

FAO (2015) Status of the World's Soil Resources (SWSR)–Main report. In: Food and agriculture organization of the united nations and intergovernmental technical panel on soils. Rome, Italy, FAO, p 650

Geist HJ, Lambin EF (2004) Dynamic causal patterns of desertification. Bioscience 54(9):817–829

Google earth imagery V 6.2.2.6613. (December 30, 2006, and December 30, 2016). North-western coastal areas, Libya. 32°50′ 18.80″ N, 13°30′31.82″ E, Eye alt 31.09 miles. Copernicus 2018. http://www.earth.google.com. Accessed 24 July 2020

GTZ (2009) Monitoring and assessing land degradation processes: selected methods and case studies. Federal Ministry for Economic Cooperation and Development, Germany

Hulme M, Kelly M (1993) Exploring the links between desertification and climate change. Environment 35:4,39–11,45

IPBES (2018) The IPBES assessment report on land degradation and restoration. In: Montanarella L, Scholes R, Brainich A (eds). Secretariat of the intergovernmental science-policy platform on biodiversity and ecosystem services, Bonn, Germany
IPCC (2001) Climate change 2001. In: Impacts, adaptation, and vulnerability. Contribution of working group II to the third assessment report of the intergovernmental panel on climate change. Cambridge University Press, New York
IPCC (2007) Climate change 2007. In: IPCC fourth assessment report working group I report "The Physical Science Basis". Cambridge University Press, New York
Lambin EF (2001) The causes of land-use and land-cover change: moving beyond the myths. Global Environ Change 11:261–269
MacDonal AM, Bonsor HC, Dochartaigh BEO, Taylor RG (2012) Quantitative maps of groundwater resources in Africa. Environ Res Lett 7024009
Mansour A, Shariff NM, Gairola S (2011) Nature and causes of land degradation and desertification in Libya: need for sustainable land management. Afr J Biotechnol 10(63):13680–13687
MEA (2005) Ecosystems and human well-being: desertification synthesis. World Resources Institute, Washington, D.C. pp. 2
NAP (2002) National action plan for combating desertification. National Committee to Combat Desertification, Ministry of Agriculture, Libya
Nwer B (2005) The application of land evaluation technique in the north-east of Libya. Doctoral thesis. Cranfield University. UK
Reynolds JF, Stafford Smith DM (2002) Do humans cause deserts? In: Reynolds JF, Stafford Smith DM (eds) Global desertification: do humans cause deserts? Berlin, Dahlem University Press, pp. 1–21
Saunders I, Young A (1983) Rates of surface process on slopes, slope retreat and denudation. Earth Surf Process Landf 8:473–501
Sevink J, Imeson AC, Verstraten JM Humus form development, and hillslope runoff, and the effects of fire and management, under Mediterranean forest in NE-Spain. CATENA 16(4–5):461–475
Thomas DSG (1997) Science and the desertification debate. J Arid Environ 37(4):599–608
UNCCD (2007) United nations convention to combat desertification. Official website of the united nations convention to combat desertification. http://www.unccd.int/
UNCED (1992) Managing fragile ecosystems: combating desertification and drought (Agenda 21, Chapter 12). United Nations
UNEP United Nations Environmental Programme (1997) The World Atlas of Desertification (eds) Middleton and Thomas D, Arnold, Great Britain. ISBN: 0340691662
World Bank (1998) World development report 1998/1999: knowledge for development. New York : Oxford University Press. World Bank
Yang J, Prince SD (2000) Remote sensing of savanna vegetation changes in eastern Zambia 1972–1989. Int J Remote Sens 21:301–332
Zurqani H, Mikhailova E, Post C, Schlautman M, Sharp J (2018) Predicting the classes and distribution of salt-affected soils in north-west Libya. Commun Soil Sci Plant Anal 49(6):689–700
Zurqani HA, Mikhailova EA, Post CJ, Schlautman MA, Elhawej AR (2019) A review of Libyan soil databases for use within an ecosystem services framework. Land 8(5):82

Dr. Bashir A. Nwer was born in Libya in 1971. He is a Ph.D. graduate from Cranfield University, UK, 2006, in the field of land evaluation using geographic information System (GIS). Dr. Nwer has great interest in the field of land degradation and combating desertification. He was Chairman of the National Committee to Combat Desertification in Libya. He was vice-chairman of the committee of the review of the implementation of UNCCD and a rapporteur of the Ninth and Tenth Session of CRIC, UNCCD. Dr. Nwer has attended UNCCD, COP5-COP11, and most of the regional meetings of UNCCD. Dr. Nwer has published a number of papers in the use of GIS and RS in land evaluation and degradation.

Dr. Hamdi A. Zurqani is one of the faculty members at the Department of Soil and Water Sciences, Faculty of Agriculture, University of Tripoli, Tripoli, Libya. Dr. Zurqani is a recognized expert as a result of his internationally acclaimed work in the areas of environmental information science, remote sensing, land evaluation, sustainability, pedology, and soil science education. He has conducted research across the world, including the United States of America and Africa. Dr. Zurqani is a distinguished soil scientist with a wide range of scientific and working experiences in Libya and abroad. He received his M.Sc. (2010) from the University of Tripoli, Tripoli, Libya and Ph.D. (2019) from Clemson University, Clemson, SC, USA. His major research and teaching activities at the University of Tripoli have focused mainly on soil genesis and classification and the environmental information sciences (Remote Sensing and Geographic Information System). He has published broadly in many journals (e.g., Nature "Scientific Reports", Geoderma; International Journal of Applied Earth Observation and Geoinformation; Journal of Geological Sciences; Land; Frontiers in Environmental Science; Communications in Soil Science and Plant Analysis; and others). Dr. Zurqani is a member of the Editorial Board for Remote Sensing (MDPI) Journal, counseling outcome and research evaluation. He also was appointed to serve as a Guest Editor for the Special Issue "Applications of Remote Sensing in Earth Observation and Geo-Information Science". In addition, Dr. Zurqani conducted peer-review for many journals including Journal of Environmental Informatics, Applied Sciences, SN Applied Sciences, Remote Sensing, Heliyon, Geosciences, Land, Water, Agronomy, Agriculture, Sustainability, Arid Land Research and Management, International Journal of Environmental Research and Public Health, Natural Hazards, and Conference of the Arabian Journal of Geosciences. He is also one of the authors of the lab manual entitled, "GIS Exercises for Natural Resource Management". Dr. Zurqani has been the recipient of numerous awards and honors: Recipient of Douglas R. Phillips Award for Graduate Students, Department of Forestry and Environmental Conservation, Clemson University, April 12, 2019; the First Place Best Judged Poster (CAFLS) at the GRADS 2019: Clemson Student Research Forum on April 4, 2019; the Second Place Poster at the 11th Clemson Biological Sciences Annual Student Symposium, April 6, 2019; the Second Place Best Judged Poster at the Clemson Student Research Forum on April 4, 2018; and the Third Place Poster at the 9th Clemson Biological Sciences Annual Student Symposium, February 25, 2017. Dr. Zurqani conducts cutting-edge research in the field of environmental information science, remote sensing, land use management/planning, change detection of landscape degradation, and geographic information system (GIS) models. He has focused on his research efforts on the development of new technologies in the field of environmental information sciences, geo-intelligence (advanced geo-information science and earth observation, machine and deep learning, and big data analytics), remote sensing, land evaluation, pedology, land use management/planning, monitoring and evaluating sustainable land management, change detection of landscape degradation, and geographic information system models.

Dr. Azalarib. S. Ali was born in Libya 1971. He is a Ph.D. graduate from Cranfield University, UK, 2018 in the field of early warning system of desertification and land degradation. Dr. Salem has got master's degree in Water Resource and a master's degree in Application of Remote Sensing in Environmental Science. Dr. Salem has published a number of papers and publications in the use of remote sensing in environment. He works as lecturer in Soil and Water Department, Faculty of Agriculture, University of Tripoli.

Index

H. A. Zurqani (ed.), *The Soils of Libya*, World Soils Book Series,
https://doi.org/10.1007/978-3-030-66368-1